Computer Architecture and Design Methodologies

Twilight zone of Moore's law is affecting computer architecture design like never before. The strongest impact on computer architecture is perhaps the move from unicore to multicore architectures, represented by commodity architectures like general purpose graphics processing units (gpgpus). Besides that, deep impact of application-specific constraints from emerging embedded applications is presenting designers with new, energy-efficient architectures like heterogeneous multi-core, accelerator-rich System-on-Chip (SoC). These effects together with the security, reliability, thermal and manufacturability challenges of nanoscale technologies are forcing computing platforms to move towards innovative solutions. Finally, the emergence of technologies beyond conventional charge-based computing has led to a series of radical new architectures and design methodologies.

The aim of this book series is to capture these diverse, emerging architectural innovations as well as the corresponding design methodologies. The scope covers the following.

- Heterogeneous multi-core SoC and their design methodology
- Domain-specific architectures and their design methodology
- Novel technology constraints, such as security, fault-tolerance and their impact on architecture design
- Novel technologies, such as resistive memory, and their impact on architecture design
- Extremely parallel architectures

More information about this series at http://www.springer.com/series/15213

Sascha Roloff · Frank Hannig ·
Jürgen Teich

Modeling and Simulation of Invasive Applications and Architectures

Springer

Sascha Roloff
Department of Computer Science
Friedrich-Alexander-Universität
Erlangen-Nürnberg
Erlangen, Bayern, Germany

Frank Hannig
Department of Computer Science
Friedrich-Alexander-Universität
Erlangen-Nürnberg
Erlangen, Bayern, Germany

Jürgen Teich
Department of Computer Science
Friedrich-Alexander-Universität
Erlangen-Nürnberg
Erlangen, Bayern, Germany

ISSN 2367-3478 ISSN 2367-3486 (electronic)
Computer Architecture and Design Methodologies
ISBN 978-981-13-8389-2 ISBN 978-981-13-8387-8 (eBook)
https://doi.org/10.1007/978-981-13-8387-8

This Springer imprint is published by the registered company Springer Nature Singapore Pte Ltd.
The registered company address is: 152 Beach Road, #21-01/04 Gateway East, Singapore 189721, Singapore

Preface

As a result of the continuous technology scaling in modern semiconductor chips, heterogeneous multi- and many-core processors have emerged in the past years as the preferred solution to exploit the performance potential of the available transistors. Traditional communication systems such as shared buses or crossbars do not fulfill the increased bandwidth requirements between the cores. For that reason *networks-on-chips* (NoCs) have been proposed as scalable and modular interconnect structure. This trend toward NoC-based *multiprocessor systems-on-chips* (MPSoCs) implementing hundreds to thousands of cores as well as hardware accelerators on a single chip leads to many different problems such as overheating, reliability, and power issues as well as resource contention. As a remedy, *invasive computing* has been proposed as a novel paradigm for designing and programming such architectures in a resource-aware and self-adaptive manner. By introducing the primitives *invade*, *infect*, and *retreat*, an application is allowed to exclusively claim resources by itself in dependence on its current computing demands and the state of the underlying resources, dynamically spread its computations for parallel execution, and deallocate them again after termination. Expected benefits of this self-organizing behavior are better resource utilization, energy efficiency, fault tolerance, and predictable execution times.

In order to study the dynamic behavior of invasive programs and to investigate a broad range of different invasive architectures without the need to have full hardware and software implementations available, fast and flexible simulation techniques are mandatory. Therefore, the first contribution of this book is the design and development of the full-system simulator InvadeSIM, which allows modeling complex invasive architectures, emulating the behavior of the invasive run-time system, and simulating function and timing of invasive applications. Existing simulation approaches do not provide proper abstraction levels, require too detailed simulation models, or are too slow for an efficient simulation of 1000-core MPSoCs. Therefore, a novel high-level simulation approach is proposed to tackle the complexity and the heterogeneity of current multi- and many-core architectures. This includes a timing approximate processor simulation approach, which is based on a discrete-event, direct execution simulation, and a timing estimation mechanism called *time warping* to scale the elapsed execution times from the host processors to the simulated target processors. Furthermore, timing models for accelerators such as *tightly coupled processor arrays* (TCPAs) as well as a contention model of concurrent accesses to the shared memory are presented. Also, an efficient NoC simulation approach is presented that allows simulating NoC communication delays equally accurate but in average much faster than a cycle-accurate simulator. Proposed are novel algorithmic and analytical techniques, which dynamically predict the transmission delays of messages considering the actual congestion in the NoC, routing information, packet lengths, and other parameters. In order to exploit multi-core host machines for fast simulations, four novel parallel discrete-event simulation techniques are proposed, which exploit the parallelism of the simulated target architecture and applications for parallel execution on the simulation host machine. An architecture description file format allows configuring and modeling various invasive architectures easily.

These efficient and flexible simulation techniques pave the way for detailed simulation-based studies on invasive programming concepts, application mapping approaches, and timing predictability. In order to guarantee real-time requirements at design time independent from dynamic workloads, so-called *hybrid application mapping* (HAM) approaches have been introduced and are the focus of the second contribution of this book. In the context of HAM approaches, the contributions of this book are furthermore (a) the design and development of a novel actor-oriented programming library called ActorX10, which allows to describe computation and communication of stream processing applications formally being the fundamental application model used in *hybrid application mapping* (HAM), (b) the development and evaluation of an efficient mapping algorithm based on backtracking to find constellations of resources fulfilling the statically determined resource constraints at run-time, and (c) a simulation-based real-world case study from the computer vision domain to demonstrate the capabilities of HAM approaches to provide predictable executions of stream processing applications on heterogeneous MPSoCs and to guarantee statically analyzed best and worst-case timing requirements on latency and throughput.

Erlangen, Germany

Sascha Roloff
Frank Hannig
Jürgen Teich

Acknowledgements

I would like to express my deep gratitude to Prof. Dr.-Ing. Jürgen Teich for his constant support and patient guidance, and his enthusiasm and attention in driving my research, the valuable scientific and technical discussions, and the opportunity to work under excellent research conditions. It is a privilege to have a supervisor that cares so much about your work and your personal development to always give nothing less but the best what you are able to provide. I would like to offer my special thanks to Dr.-Ing. Frank Hannig, who always supported me in any regard with inspiring and motivating advices and brought not only fun into the office, but also energizing coffee after long nights of writing. Furthermore, I thank all the members of the Department of Computer Science that helped me to advance my research and to finish this work. Especially Stefan Wildermann, Alexandru Tanase, Oliver Reiche, Tobias Schwarzer, and Éricles Sousa deserve my thanks. I express my gratitude to my patient girlfriend, family, and friends, who always have been my rock in the waves during the ups and downs of my research. As Supervisor, I also thank my students for helping me to keep things in perspective. Finally, I would like to thank the German Research Foundation for providing the research funding and the University of Erlangen-Nürnberg for giving me the opportunity to attend conferences and meet a lot of interesting people around the world.

Contents

Acronyms

APGAS	Asynchronous Partitioned Global Address Space (Sect. 2.2.1)
API	Application Programming Interface (Sect. 3.3.4)
BE	Best Effort (Sect. 2.1.1.3)
C*i*C	Core *i*-let Controller (Sect. 2.1.1)
CPU	Central Processing Unit (Chap. 1)
CSP	Constraint Satisfaction Problem (Sect. 6.1.4)
DES	Discrete-Event Simulation (Sect. 1.1)
DoP	Degree of Parallelism (Sect. 2.1)
DPM	Dynamic Power Management (Sect. 3.3.4)
DSE	Design Space Exploration (Chap. 5)
DVFS	Dynamic Voltage and Frequency Scaling (Sect. 3.3.4)
E3S	Embedded System Synthesis Benchmarks Suite (Sect. 6.3.3)
ESL	Electronic System Level (Chap. 3)
FFT	Fast Fourier Transform (Sect. 3.3.5)
FIFO	First-in, First-out (Sect. 2.2.2.3)
FIR	Finite Impulse Response (Sect. 2.1.1.2)
FLIT	Flow Control Digit (Sect. 4.1)
FPGA	Field-Programmable Gate Array (Sect. 2.1.1.1)
FPS	Frames Per Second (Sect. 6.1.3)
FPU	Floating-Point Unit (Sect. 2.1.2.2)
FSM	Finite-State Machine (Sect. 6.2.1)
GALS	Globally Asynchronous, Locally Synchronous (Sect. 2.2.1)
GPU	Graphics Processing Unit (Chap. 1)
GS	Guaranteed Service (Sect. 2.1.1.3)
HAM	Hybrid Application Mapping (See Preface)
HPC	High-Performance Computing (Chap. 1)
ICT	Irreversible Color Transformation (Sect. 2.1.2.4)
ID	Identification (Sect. 2.2.4)

i-NoC	Invasive Network-on-Chip (Sect. 2.1.1)
IP	Internet Protocol (Sect. 2.2.3)
IPC	Inter-Process Communication (Sect. 2.2.3)
*i*RTSS	Invasive Run-Time Support System (Sect. 2.1.3)
ISA	Instruction Set Architecture (Sect. 2.1.1.1)
MIC	Many Integrated Core (Chap. 1)
MIPS	Million Instructions Per Second (Sect. 5.1.3)
MoC	Model of Computation (Sect. 6.2.1)
MPI	Message Passing Interface (Sect. 2.2)
MPPA	Massively Parallel Processor Array (Chap. 1)
MPSoC	Multiprocessor System-on-Chip (Chap. 1)
MSE	Mean Squared Error (Sect. 3.3.1)
NA	Network Adapter (Sect. 2.1.1)
NI	Network Interface (Sect. 6.3.1)
NoC	Network-on-Chip (See Preface)
NUCC	Non-Uniform Cluster Computing (Sect. 2.2)
OpenMP	Open Multi-Processing (Sect. 2.2)
PDE	Partial Differential Equation (Sect. 6.2.3.2)
PDES	Parallel Discrete-Event Simulation (Sect. 5.1.1)
PE	Processing Element (Sect. 2.1.1.2)
PERCS	Productive, Easy-to-use, Reliable Computer System (Sect. 2.2)
PFH	Probability of Failure per Hour (Sect. 6.1.3)
PGAS	Partitioned Global Address Space (Sect. 2.1.2)
PMU	Performance Monitoring Unit (Sect. 3.3.1)
POSIX	Portable Operating System Interface (Sect. 2.2.2.2)
QoS	Quality of Service (Sect. 2.1.1.3)
RANSAC	Random Sample Consensus (Sect. 6.2.3.1)
RISC	Reduced Instruction Set Computer (Chap. 1)
RPC	Remote Procedure Call (Sect. 2.1.3)
rr	Round Robin (Sect. 4.1)
RRE	Run-time Requirement Enforcement (Sect. 7.2)
SCC	Single-chip Cloud Computer (Chap. 1)
SDF	Synchronous Data Flow (Sect. 6.2.3.1)
SIFT	Scale-Invariant Feature Transform (Sect. 6.2.3.1)
SIMD	Single Instruction, Multiple Date (Sect. 2.1.2.4)
SL	Service Level (Sect. 4.1)
SMP	Symmetric Multiprocessing (Sect. 2.2)
SMT	Simultaneous Multi-Threading (Sect. 5.1.3)
SPMD	Single Program, Multiple Data (Sect. 2.2.1)
TCPA	Tightly Coupled Processor Array (Sect. 1.1)
TDMA	Time-Division Multiple Access (Sect. 4.1)
TLM	Tile-Local Memory (Sect. 2.1.1)
TS	Time Slot (Sect. 2.1.1.3)

VC	Virtual Channel (Sect. 2.1.1.3)
VLIW	Very Long Instruction Word (Chap. 1)
WRR	Weighted Round Robin (Sect. 2.1.1.3)

Chapter 1
Introduction

One of the most important drivers of technological advancements in microprocessors over the past half century was the continuous decrease in transistor size according to Moore's law [1]. It states that the number of transistors per chip will double approximately every two years, which corresponds to an exponential growth. Figure 1.1 shows that this prediction holds since more than 45 years until today and it most probably will continue for the next few years before physical limits are reached. This continuous availability of new transistors encouraged major revolutions in computer architecture with the most important one probably being the paradigm shift from *single-core* towards *multi-* and *many-core* processors.

For decades, chip manufacturers increased the performance of new processor generations by driving the clock frequency as a result of technology scaling and by optimizing the microarchitecture using techniques such as pipelining, branch prediction, multiple execution units, out-of-order execution, etc. Single-threaded applications enjoyed free and regular performance gains without any changes. However, exploiting these traditional approaches for boosting processor performance was limited. Around 2005, chip manufacturers faced physical issues in further increasing the operational frequency of the processors, since it would result in a huge power consumption and an inevitable overheating, which could not be dissipated without impracticably high cooling effort (see Fig. 1.1). Also, increasing the instruction throughput by architecture improvements reached its limits. Thus, in order to maintain performance gains and to keep Moore's law alive, processor manufacturers introduced *hyper-threading* and *multi-core* architectures. Here, the available transistors on a chip are used to replicate the internal execution units of a *central processing unit* (CPU) into cores, each executing a separate and independent instruction stream. Now, the performance gain is achieved through the exploitation of different levels of parallelism instead of an increased frequency. This radical change in hardware technology led to a fundamental turning point in software development, since existing single-threaded applications must be restructured toward parallel execution in order to take advantage of new processor generations. As a result, parallel programming has left its niche of *high-performance computing* (HPC) and is going to become

S. Roloff et al., *Modeling and Simulation of Invasive Applications and Architectures*, Computer Architecture and Design Methodologies,
https://doi.org/10.1007/978-981-13-8387-8_1

mainstream for software development targeting desktop computers, embedded or mobile devices, or even automotive systems. This paradigm shift from single-core to multi-core processors and its significant consequences on software development was described by Herb Sutter already 2005 in his famous article "The Free Lunch is Over" [2, 3].

In Fig. 1.1 can be seen that the number of cores on a chip increased since 2005 almost similar to the number of transistors. Careful predictions state that they will double approximately every 18 to 24 months corresponding to Moore's predictions for transistors. Chip manufacturers exploited this potential not only to build homogeneous dual, quad, or octa core processors (e.g., Intel Core Duo, Core i7, or Xeon E7 processor), where each core is functionally as well as structurally equivalent and able to execute any system task, but also to design heterogeneous multi-core processors, where a mixture of powerful and weaker cores are used to improve energy efficiency (e.g., ARM big.LITTLE architecture or IBM Cell processor). Furthermore, specialized hardware accelerators are used for specific tasks such as signal processing, network communication, cryptography, or artificial intelligence. Together with the cores of a multi-core processor, on-chip memory, and an interconnect, they are integrated into one complex *multiprocessor system-on-chip* (MPSoC). The communication infrastructure is often realized by a so-called *network-on-chip* (NoC), which provides a regular structure of simple building blocks and scales up to thousands of cores compared to classical interconnects like single arbitrated busses or crossbars. Such heterogeneous many-core architectures enable high flexibility in resource usage, increased energy efficiency as well as high performance computing potential

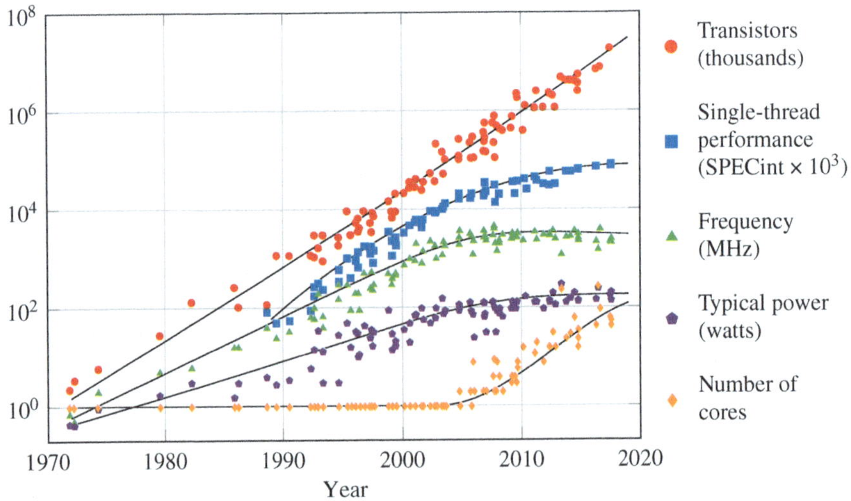

Fig. 1.1 Microprocessor trend data for the last 45 years. While frequency and power stagnates since 2005, the number of transistors still increases exponentially and leads to a similar growth of the core count since 2005. Data taken from [4] and supplemented by trendlines

in comparison to single-core processors. They have emerged as the leading solution to power and scalability issues and might also help to circumvent the problem of *dark silicon*. This term describes the phenomenon that only a small fraction of the available chip area can be used at full frequency at the same time. The remaining area must be unclocked or at least underclocked to stay within the power budget of the microprocessor, otherwise it would cause overheating or even physical damage. This so-called *utilization wall* is getting exponentially worse with each new process generation.

Consequently, due to the benefits of such many-core architectures, processor manufacturers were encouraged to integrate more and more processor cores into a single chip. Already now, chips with more than 100 cores are existing and in the near future, it is expected to have up to 1,000 cores on a chip. Current microarchitectures of *graphics processing units* (GPUs) such as Nvidia's Volta architecture already contain up to 5,120 shader cores. However, these cores are less powerful and operate in a completely different way compared to the cores of a multi-core processor. They are highly optimized for data parallel applications such as computer graphics, image processing, linear algebra, or neuronal networks and are limited in their use for general-purpose computing. Thus, GPU cores are not counted as cores as introduced before. An example of a current x86 many-core architecture is Intel's Xeon Phi co-processor, also referred to as Intel's *many integrated core* (MIC) architecture. The latest generation of this series (codename *Knights Mill*) contains up to 72 low-power Atom cores and is manufactured in 14 nm process technology. It is based on the architectures of the *Larrabee* GPU design and the *single-chip cloud computer* (SCC), which includes 48 Pentium cores connected through a 4×6 2D-mesh. Another example is the Tilera (acquired by Mellanox) TILE-Gx72 processor, which is optimized for networking, multimedia, and cloud computing applications and includes 72 identical processor cores interconnected with a mesh-based on-chip network. Kalray's *massively parallel processor array* (MPPA) is another popular representative of a high throughput, low latency many-core architecture. The second generation (codename *Bostan*) is particularly designed for time-critical applications and already contains 256 time predictable and energy efficient *very long instruction word* (VLIW) cores. Moreover, many-core processors are also extensively used in current *TOP500* supercomputers to provide extraordinary computing power. For example, the current number one supercomputer Sunway *TaihuLight* incorporates 40,960 64-bit *reduced instruction set computer* (RISC) many-core processors each containing 256 processing cores as well as 4 system management cores, which results in a remarkable total amount of 10,649,600 cores. This supercomputer does not only outperform clusters that include more cores in total yet less cores per processor, but also systems that rely on GPUs for acceleration. It is even listed among the top twenty most energy-efficient supercomputers from the *Green500* list.

Above mentioned architectures have already become an integral part of today's embedded systems driving the technological progression of devices from different domains such as mobile communication, automotive, or avionics. However, they introduce comprehensive challenges for programmers as well as designers of such systems, especially if applications with different levels mixed criticality or real-time constraints have to be considered. For instance, an automotive emergency breaking or pedestrian detection system has a much higher level of criticality according to errors in computed data compared to a multimedia application. Applications have to deal with heterogeneity, extreme concurrency, and unreliable hardware. Multiple levels of parallelism have to be exploited (e.g., instruction, loop, data, and task level) to benefit from an increased amount of cores. Static and centralized management concepts for controlling the allocation of the hardware resources to applications will not scale anymore. Decreasing feature sizes drastically affect the rates of temporary and permanent faults and restricted power budgets lead to the problem of dark silicon. Furthermore, guaranteed execution qualities for applications on embedded multi-core systems is of high importance.

As a remedy, the idea of *Invasive Computing* has been proposed by Teich [5] as novel paradigm for designing and programming such architectures in a resource-aware and self-adaptive manner while considering requirements on non-functional execution properties. The main idea of invasive computing is to introduce resource-aware programming in a sense that a programmer may explicitly specify and argue about resource requirements desired or required in different execution phases. In a phase of *invasion*, the application requests the runtime system to exclusively allocate a set of resources including processor cores, memory, or communication channels. In a subsequent *infect* phase, the workload is spread for parallel execution on the obtained claim of resources. Finally, if an application is terminated or the degree of parallelism has been reduced, a *retreat* phase deallocates either all or a subset of the claimed set of resources and the application resumes an execution of lowered parallelism. Expected benefits of this self-organizing behavior are better resource utilization, energy efficiency, fault tolerance, and predictable execution times. To efficiently support the operations for invasion, infect, and retreat, novel programming constructs, languages, runtime systems, and architecture concepts are required.

1.1 Goals and Contributions

The goal of this book is to propose novel simulation techniques and their implementation that enable the validation and variants' exploration of all essential features of invasive computing in order to guide research directions without the need to have full hardware or software implementations available and then to model invasive applications as actor graphs and use simulation for analyzing the predictable execution behavior and supporting mapping approaches. Therefore, the first contribution of this book is the design and development of the full-system simulator InvadeSIM, which models and simulates all layers of an invasive computing platform. An overview of these layers and the corresponding contributions are depicted in Fig. 1.2a. The main focus of InvadeSIM is on the parallel architecture simulation including processor cores, accelerators, memory accesses, and the the network-on-chip. Moreover, it includes an emulation of the invasive runtime system behavior, which bridges invasive applications[1] and the simulated invasive architectures. For a better understanding of the characteristics of invasive program behavior, InvadeSIM provides a visualization layer, which allows visualizing executions either statically by Pajé traces or dynamically in real time by the tool InvadeVIEW. The second contribution of this book is the design and development of a novel actor-oriented programming library called ActorX10, which allows modeling invasive applications as actor graphs formally and analyzing predictable execution behavior. Such applications are used as application model in hybrid mapping approaches, which aim to map applications on multi-core architectures guaranteeing certain execution properties such as a maximum execution time by a combination of design time analysis and run time embedding. An overview of hybrid application mapping and the corresponding contributions are depicted in Fig. 1.2b.

InvadeSIM

In order to study the dynamic behavior of invasive programs and to investigate a broad range of different invasive architectures without the need for a full hardware and software implementation, fast and flexible simulation techniques are mandatory. Therefore, one main contribution of this book is the design and development of the full-system simulator InvadeSIM, which allows modeling complex invasive architectures, emulating the execution behavior of the invasive runtime system, and simulating function and timing of invasive parallel applications apart from utilization, efficiency, and competition. A novel high-level processor simulation approach based on direct-execution and a linear timing estimation model is proposed that tackles the complexity and the heterogeneity of current multi and many-core architectures. Furthermore, a *discrete-event simulation* (DES) framework has been developed that allows integrating and synchronizing different simulation tasks such as software or hardware simulations. A shared-memory bandwidth congestion model and a dynamic frequency adaption technique have been also developed as part of InvadeSIM. Besides processor simulation, different timing simulation models for hardware accelerators such as *tightly-coupled processor array* (TCPA) or *i*-Core have been developed.

Hybrid Network-on-Chip Simulation

Modern multi- and many-core architectures heavily use communication infrastructures such as NoCs in order to meet the increased bandwidth demands between the processors and for reasons of scalability. Here, an efficient hybrid NoC simulation approach that allows simulating communication delays equally accurate but in average much faster than on a cycle-by-cycle basis is proposed. This includes novel algorithmic and analytical techniques, which dynamically predict the transmission delays of messages considering the actual congestion in the NoC, routing information, packet lengths, and other parameters. According to these predictions, the simulation time may be automatically advanced in many cases, which drastically

[1] Invasive applications are written in the programming language X10 and make use of the invasive programming library InvadeX10.

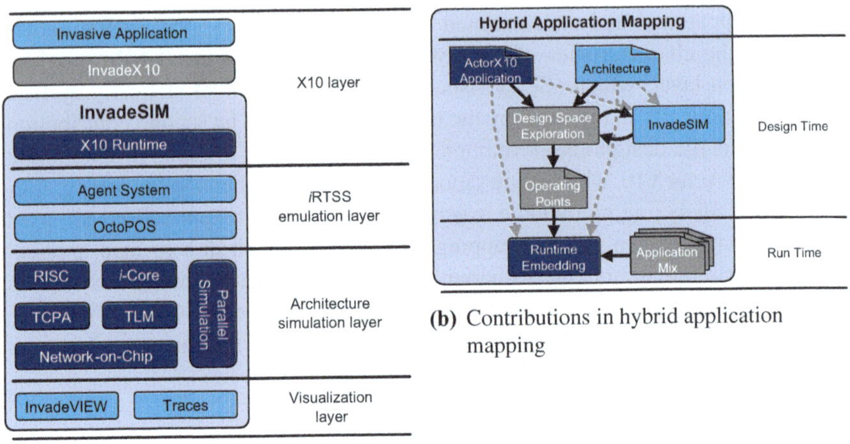

(a) Contributions in InvadeSIM

(b) Contributions in hybrid application mapping

Fig. 1.2 Overview of the contributions of this book in InvadeSIM and hybrid application mapping. Dark blue highlights the major contributions, which represent the main content of this book. Light blue are additional contributions, but are not part of the corresponding chapters. Gray are contributions from others

reduces the number of cycles the NoC simulator has to process. Furthermore, the integration of the proposed NoC simulation technique into the simulator InvadeSIM will be shown.

Parallel Simulation

The simulation of modern parallel multi-core and many-core systems including parallel software and hardware containing hundreds to thousands of heterogeneous cores can be very costly and time consuming. Fast functional and timing simulation techniques are required that exploit the parallelism of multi-core host machine. Here, four novel parallel discrete-event simulation techniques for InvadeSIM are proposed, which exploit the parallelism of the simulated target architectures and applications for parallel simulation on the host machine. In order to achieve this, correct synchronization and activation of the host threads is required and solved differently by the four proposed parallelization techniques. Furthermore, parallel simulation allows evaluating different architectural design choices such as number of tiles, internal tile structure and selection of cores within each tile. Here, we present case studies regarding performance and costs trade-offs of different heterogeneous invasive architecture variants.

ActorX10 and Run-Time Application Embedding

In order to guarantee real-time requirements at design time independent from dynamic workloads, so-called *hybrid application mapping* (HAM) approaches have been introduced and are the focus of the second contribution of this book. In the context of *hybrid application mapping* (HAM) approaches, the design and implementation of a novel actor-oriented programming library called ActorX10 is presented, which allows describing computation and communication of stream processing applications formally. In order to find constellations of resources fulfilling statically determined resource constraints at run time, the implementation and evaluation of an efficient problem-specific mapping algorithm based on backtracking is presented. Furthermore, we present simulation-based real-world case studies to demonstrate the capabilities of invasive computing to provide predictable executions of stream processing applications on heterogeneous MPSoCs and to guarantee statically analyzed best and worst-case timing requirements on latency and throughput.

1.2 Organization of the Book

In Chap. 2, we introduce the basic principles of invasive computing including the considered type of tiled architectures and the novel language constructs. Furthermore, a tutorial on the parallel programming language X10 is provided that introduces its programming model, language constructs for concurrency and communication, and runtime implementation. In Chap. 3, our simulation framework InvadeSIM is presented including the novel direct-execution simulation approach and the DES framework. Furthermore, timing simulation models for the TCPA and the *i*-Core are presented. Chapter 4 presents our novel hybrid NoC-simulation approach and how it is integrated into InvadeSIM. Our novel parallel simulation techniques and architecture evaluations are presented in Chap. 5. In Sect. 6.2, we introduce basic definitions on predictability, present ActorX10, our novel run-time mapping algorithm, and our predictability case studies.

References

1. G.E. Moore, Cramming more components onto integrated circuits. Electronics **38**(8), 114–117 (1965). ISSN: 0013-5070
2. H. Sutter, J. Larus, Software and the concurrency revolution. Queue-Multiprocessors **3**(7), 54–62 (2005). ISSN: 1542-7730. https://doi.org/10.1145/1095408.1095421
3. H. Sutter, The free lunch is over: a fundamental turn toward concurrency in software. Dobb's J. **30**(3), 202–210 (2005), http://www.gotw.ca/publications/concurrency-ddj.htm
4. K. Rupp, 42 Years of Microprocessor Trend Data (2018), https://www.karlrupp.net/2018/02/42-years-of-microprocessor-trend-data
5. J. Teich, Invasive algorithms and architectures. IT-Inf. Technol. **50**(5), 300–310 (2008). ISSN: 1611-2776. https://doi.org/10.1524/itit.2008.0499

Chapter 2
Fundamentals

Highly parallel many-core architectures as introduced in Chap. 1 incorporate already hundreds of cores on a single chip and the number will increase in the future. This huge amount of available hardware resources introduce major challenges in programming and designing such architectures, especially when considering an increasing rate of temporary and permanent faults due to decreasing feature sizes. Conventional concepts of centralized resource management strategies as well as static parallelization approaches will not scale anymore. As a remedy, a novel programming paradigm called *Invasive Computing* has been proposed by Teich [1] to tackle current and upcoming challenges in programming and designing massively-parallel MPSoCs. This chapter provides an introduction into the basic principles of invasive computing and the realization of the concepts in hardware and software, which serve as simulation target for InvadeSIM. Furthermore, the parallel programming language X10 as fundamental basis for realizing the ideas of invasive computing and for all implementations in this book including the full-system simulator InvadeSIM and the actor-oriented programming library ActorX10 is presented.

2.1 Invasive Computing

Parallel computing is not restricted anymore to high-performance computing, but it becomes more and more important also in embedded systems as complex multiprocessor systems-on-chips are increasingly used. However, using conventional methods for programming and designing such systems has major shortcomings and leads to various challenges in the future. Among others, these are adaptivity, reliability, scalability, programmability, and physical constraints as stated in [2]. Important research questions are:

S. Roloff et al., *Modeling and Simulation of Invasive Applications and Architectures*, Computer Architecture and Design Methodologies,
https://doi.org/10.1007/978-981-13-8387-8_2

- How are programs mapped onto 1,000 cores to exploit the underlying parallelism of the hardware?
- Can programs be specified to efficiently run without changes on 1, 2, or N cores?
- How can different computing requirements of applications be taken into account and resources dynamically distributed to the applications?
- How can physical constraints such as maximum heat or power dissipation be ensured?
- Can programs be written in a way to tolerate and compensate faults caused by degradation effects?

To answer these questions and to tackle the mentioned challenges, a novel parallel programming paradigm called *Invasive Computing*[1] has been proposed by Teich [1] and is summarized in the overview article [2]. Novel concepts of resource-aware programming for managing parallel applications running on 1,000-core MPSoCs have been introduced. Programs are allowed to manage and coordinate the usage and allocation of processing, communication, and memory resources by themselves in dependence on their current computing demands and the reservation and physical state (e.g., temperature, power, faultiness, etc.) of the resources. In [2], Teich et al. denote this self-organizing program behavior as *invasive programming* and define it as follows:

Definition 2.1 (*Invasive Programming* [2]) "Invasive Programming denotes the capability of a program running on a parallel computer to request and temporarily claim processor, communication and memory resources in the neighbourhood of its actual computing environment, to then execute in parallel the given program using these claimed resources, and to be capable to subsequently free these resources again."

Through the concepts of invasive computing, an application (program)[2] is able to spread its computations for parallel execution on a set of claimed resources based on its actual computing demands and on the availability and state of the resources. For execution phases of reduced degree of application parallelism, occupied resources may be freed by the application to make them available for other applications. Applied appropriately, these principles may drastically increase resource utilization and thus, computational efficiency. In [3], the authors could verify the expected benefits of invasive computing by carefully analyzing the introduced overheads and the resulting resource efficiency compared to statically mapped applications. They showed that invasive programs could reach a significantly higher resource efficiency of almost 100% compared to 32% of non-invasive programs, while achieving similar speedups. These results clearly show the benefits of resource-aware programming compared to conventional programming concepts due to the fact that applications only claim resources, when needed and leave them available to other applications, after they have

[1]Invasive Computing is the title as well as the research focus of the Transregional Collaborative Research Center 89 (http://www.invasic.de).

[2]In the following, we use application and parallel application program synonymously.

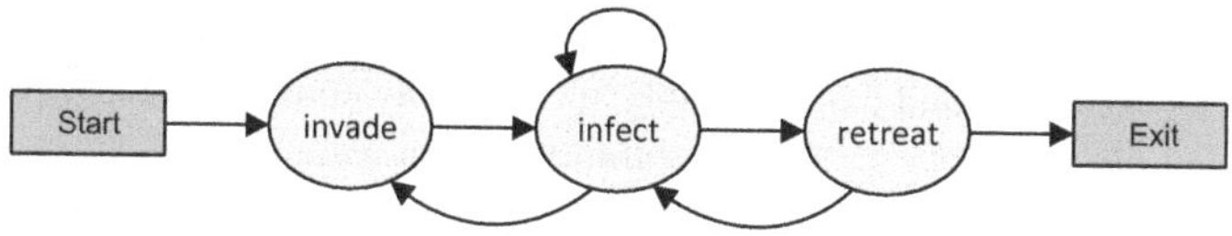

Fig. 2.1 Example of execution phases of an invasive program (adapted from [4])

used and freed them again. Furthermore, applications are able to adapt themselves to a varying amount of resources as well as to different types of resources or a changing physical state. For instance, an application may exploit a higher *degree of parallelism* (DoP), if more computing resources become available or a specially tailored algorithm might be executed for an accelerator. Also, applications may react by retreating from resources, if they become unreliable or overheated as presented in a case study in [4]. In [5], the authors explored different runtime tradeoffs regarding quality and throughput using invasive computing concepts. For example, using different algorithms for a different amount of resources may keep the resulting throughput equal, but varies in quality. In turn, using the same algorithm may keep the quality equal, but varies in throughput, if the number of resources is changed. Moreover, invasive computing allows reserving computation as well as communication resources exclusively. This isolates the applications from each other and therefore increases the predictability of non-functional execution properties such as latency, throughput, or power consumption on multi-core architectures as shown in [6, 7]. Resource management is realized in a decentralized manner, which is essential for systems including up to 1,000 or more cores, otherwise it would not be scalable. Supporting this paradigm on self-adaptive and resource-aware programming requires radical changes in hardware and software. All aspects of system design are affected including research in languages, compilers, operating systems as well as architectures.

The chart depicted in Fig. 2.1 shows typical phases and transitions between these phases during the execution of an invasive program. After starting, an invasive application expresses its demands for resources by so-called *constraints* and provides them to the runtime system in a phase called *invasion* to requests a certain number of cores or other resources for its parallel execution. A so-called *claim* is constructed, which denotes the set of resources that has been reserved for the application during the *invade* command. Afterwards, *infect* is used to start the parallel computations on the assigned claim. The entities of code that are copied to the claimed cores for subsequent parallel execution are called *i*-lets (standing for *invasive*-let[3]). The execution of *i*-lets itself is triggered by the infect command. Once the execution on all cores is finished, the number of cores inside the claim can be altered by again calling invade (*reinvade*) or *retreat* to either expand or shrink the application's claim. Alternatively, if the degree of parallelism does not change, it is also feasible to dispatch a different program onto the same set of resources by issuing another call to infect (*reinfect*). A retreat call can either free all resources from the claim or just a subset (*partial*

[3]This conception goes back to the notion of a *servlet*, which is a (Java) application program snippet targeted for execution within a web server.

retreat). If no computing resources are left for further executions of the program, it terminates its execution and exits. Notably, a claim may contain not only processing resources, but also memory as well as communication resources.

Different types of applications may profit from the introduced invasive programming concepts. In [3], applications with phases of a different dynamic degree of parallelism have been analyzed. For each phase, an application explicitly requests more or retreats from resources based on its current degree of parallelism. Once requested, resources are not returned during a phase, even if another application would profit more from additional resources. Thus, so-called *malleable* applications [8, 9] are studied in invasive computing. They are able to dynamically adapt their degree of parallelism in response to external resource requests. Instead of explicitly requesting resources, a malleable application is typically equipped with a performance hint (e.g., a speedup curve, which defines the gain in performance with respect to the number of allocated cores) and the runtime system is responsible to determine a resource allocation, which maximizes the overall performance. In [10], an agent-based resource management is presented, which decentrally solves this optimization problem. Furthermore, the authors of [11] could prove through game theoretical analyses that their presented core allocation scheme only based on the local exchange of information will always result in an optimal allocation. Also, applications with real-time constraints may profit from invasive computing concepts. They have to be represented as *static task graphs* and are analyzed at design time. Invasion is then performed on a graph structure instead on the resource constraints of a dynamically changing degree of parallelism.

Invasive computing can solve many problems of massively-parallel applications running on future MPSoC platforms by providing and porting principles of *self-organization* and *resource-awareness* into reconfigurable architectures, integrating 1,000 and more processor cores on a single chip. Another objective is to increase the lifetime or to encompass aging effects of future sub-micron technology by avoiding stressing the hardware too much. The major advantages invasive computing offers can thus be summarized as follows:

- Gain in computational efficiency and performance [3, 11],
- Application-level fault-detection and fault-tolerance [12, 13],
- Self-adaptive power control and management [14, 15], and
- Self-optimization of resource utilization [16, 17].

2.1.1 Invasive Tiled Architectures

In order to realize the idea of self-adaptation and resource-aware programming, not only new language constructs, but also architectural innovations to efficiently support the three basic primitives invade, infect, retreat on architecture level are required. This includes concepts for dynamic reconfiguration of processor, communication, and memory resources. Thus, in this section, we introduce the type of architectures studied in invasive computing and considered as simulation target for InvadeSIM.

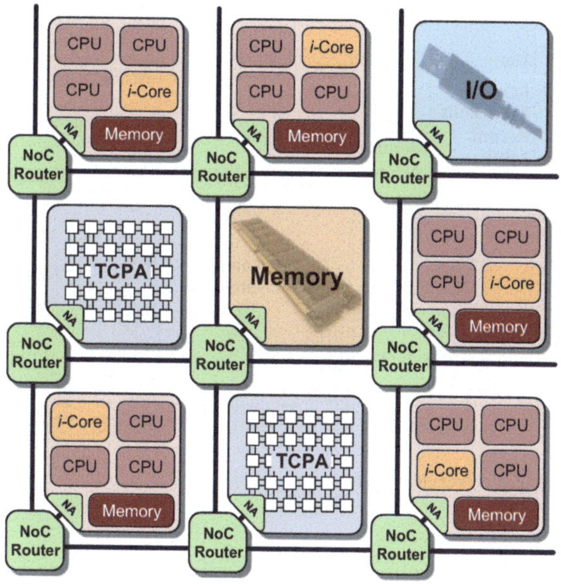

Fig. 2.2 Generic invasive tiled architecture including different types of processing resources such as loosely-coupled RISC cores, *i*-cores as well as TCPAs arranged in tiles with local memory and connected via the *i*-NoC [18]

An example of an instance of an invasive architecture is depicted in Fig. 2.2. Considered are heterogeneous tiled MPSoCs, where the available processing resources are organized into so-called tiles each including a local memory. A generic invasive tiled architecture may consist of several compute tiles, memory tiles, tiles with invasive-specific accelerators, and I/O tiles. Compute tiles are considered to be heterogeneous to support different granularities of application parallelism (e.g., instruction, loop, or thread-level parallelism). They may consist of (a) a set of loosely-coupled standard RISC cores, (b) a special-purpose RISC core, which provides dynamically reconfigurable instruction set extensions (*i*-core), or (c) a TCPA, which provides high-speed and power-efficient loop acceleration for compute-intensive applications such as signal and image processing. Each compute tile also contains a *tile-local memory* (TLM) for local data exchange, a special hardware component for dynamic thread assignment: *Core* i-*let controller* (C*i*C), and an internal bus-structure for local communication. The tiles are connected by a dynamically configurable *invasive network-on-chip* (*i*-NoC) and each tile is equipped with a *network adapter* (NA) to handle remote communication over the *i*-NoC. Additional memory tiles host larger on-chip memories and are globally accessible. I/O tiles provide interfaces to access peripherals. More details about invasive architectures can be found in [19].

2.1.1.1 The Invasive Core

One special-purpose core used in invasive computing is the runtime adaptive *invasive core* (*i*-core) [20]. It is based on a standard RISC core (SPARC V8 LEON3) and extends its micro-architecture by runtime adaptable components to support inva-

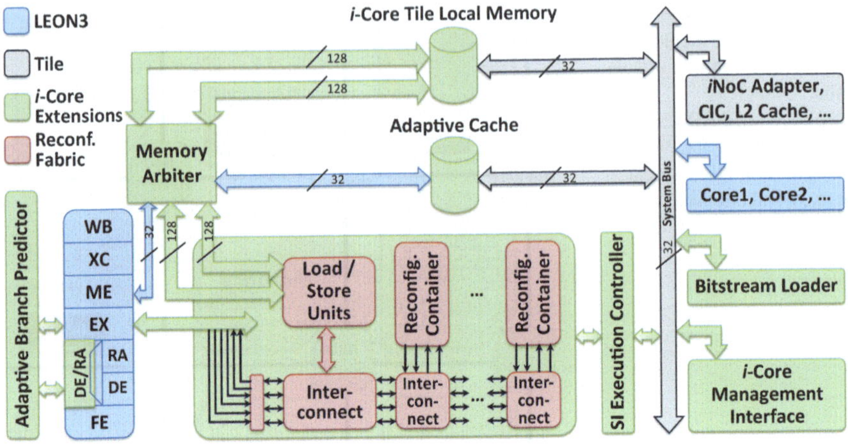

Fig. 2.3 Overview of the micro-architectural structure of the *i*-Core [20]

sive programming concepts on micro-architecture level (see Fig. 2.3). For example, the cache associativity, the pipeline stages, or the branch prediction unit can be adapted at run time, when requested by an application. Furthermore, the *i*-core allows adapting its *instruction set architecture* (ISA) at run time. Besides the regular instructions of the SPARC V8 core, it provides special instructions, which can be activated on demand at run time by loading different hardware accelerators into a so-called *reconfigurable fabric*. Thus, the *i*-Core is able to accelerate certain program phases by special instructions, but also to execute regular LEON3 programs, if no accelerator is loaded. The execution of a single special instruction typically replaces multiple regular instructions, why it increases not only performance, but also resource efficiency. Partial reconfiguration technology of *field programmable gate arrays* (FPGAs) allows this dynamic adaption and the integration of hardware accelerators into the micro-architecture of the *i*-core. Special instructions are used to accelerate compute-intensive applications such as cryptography or video compression. Speedups of up to 22 compared to a regular LEON3 core execution for an *H.264 encoder* have been reported [19].

2.1.1.2 Tightly-Coupled Processor Arrays

A special class of hardware accelerators studied in invasive computing are TCPAs [21]. They are specifically designed to accelerate computationally intensive loop programs such as one-dimensional signal processing applications, e.g., *finite impulse response* (FIR) filters or two-dimensional image processing applications, e.g., algorithms for edge detection or optical flow by jointly exploiting instruction, loop, and bit-level parallelism of such programs. Compared to general-purpose processors, they achieve much higher throughput, provide better energy efficiency, and offer a fully timing predictable execution. An exemplary tile configuration including a

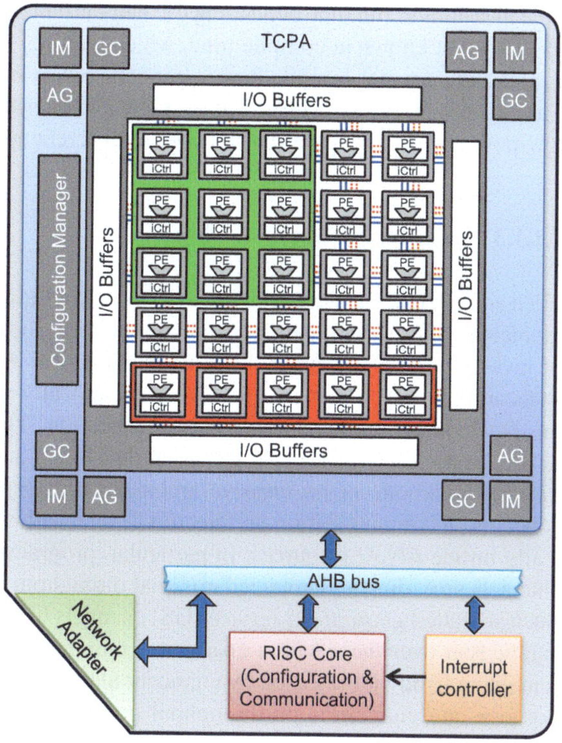

Fig. 2.4 Example configuration of a TCPA tile. A TCPA is composed of its core array structure surrounded by I/O buffers, units for execution management and configuration: Global controller (GC), address generator (AG), and configuration manager (CM), and units for supporting invasion: Invasion manager (IM) and invasion controller (iCtrl) in each PE [21]

TCPA is depicted in Fig. 2.4. Each TCPA tile always includes a regular RISC core, which controls the TCPA execution by different driver commands and can be synchronized with the TCPA using an interrupt controller. The core structure of a TCPA is an array of tightly-coupled VLIW *processing elements* (PEs) arranged in a grid layout. Each PE has its own instruction memory and can be programmed independently of each other. The type and number of functional units (e.g., adder, multiplier, shift and logical operations) inside a PE can be configured at design time as well as the number of rows and columns of the whole array. Additional hardware units (global controller, address generator, and configuration manager) are used to manage the execution and configuration of the TCPA. PEs only communicate to their direct neighbors, via point-to-point connections, which allows data exchange within a couple of cycles. They are not connected to a shared main memory. Instead, to feed the PEs with data and to store produced results, the array is surrounded by buffers, which are accessed by the border PEs at the corresponding cardinal direction. Furthermore, TCPAs also support the concepts of resource-aware programming on architecture level. The available PEs of a TCPA may be used by different applications simultaneously as illustrated in Fig. 2.4 by the areas marked in green and red. Thus, a TCPA includes hardware units (invasion manager and invasion controller in each PE) that allow an application to invade a certain area of the available PEs in competition to the requests of other applications. This may result in a different number of PEs for multiple invasions at run time depending on their current availability. Hence, the number of PEs is not known at compile time, which requires symbolic parallelization techniques and a special compile flow to generate binary code for loop-based programs. The reconfigurable structure of the TCPA allows synthesizing highly customized low-power array configurations tailored to different application domains easily.

2.1.1.3 Invasive Network-on-Chip

Tiled architectures as used in invasive computing require a powerful and flexible communication infrastructure between the tiles to serve the high bandwidth requirements of invasive streaming and multimedia applications. Additionally, invasive applications are not only considered to have a dynamically varying DOP, but also changing bandwidth requirements during an execution. Therefore, a so-called invasive network-on-chip (*i*-NoC) [18] has been developed and is used as central communication backbone of an invasive architecture. It supports invasive programming concepts on communication-architecture level by allowing an application to dynamically invade *i*-NoC resources. In particular, programs can request communication channels providing a guaranteed minimal throughput as well as a maximal latency. Such so-called *guaranteed service* (GS) channels have to be established before data can be sent over them with a guaranteed minimal communication bandwidth. The remaining bandwidth is used by connectionless *best effort* (BE) traffic, which cannot give any guarantees on throughput or latency. Both traffic types can be used simultaneously as illustrated in Fig. 2.5. Each tile has its own router, which can be addressed by any package that is sent over the *i*-NoC. Tile-router communication is achieved through an NA. A router is composed of four ports in all cardinal directions

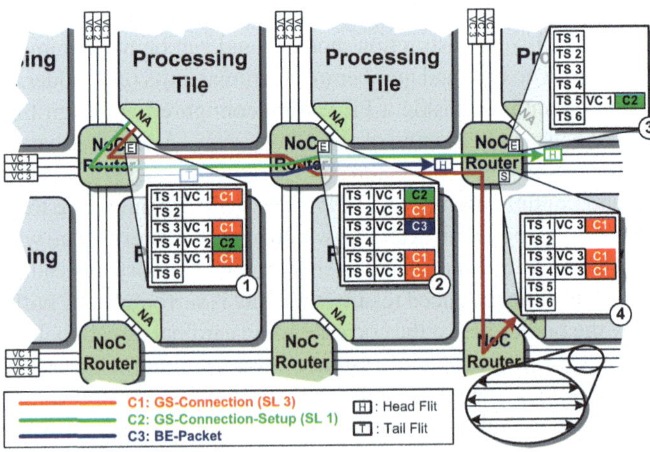

Fig. 2.5 Structural overview of the invasive network-on-chip including a communication scenario with three active connections [22]

as well as one port for the local communication with the NA. Each port has a sending (output) and receiving (input) part. The dynamic allocation of *i*-NoC resources and the guaranteed bandwidth communication is achieved through a combination of dynamic *virtual channels* (VCs) and *weighted round-robin* (WRR) scheduling. A VC is represented by the reservation of a buffer at each input port and a certain amount of *time slots* (TSs) at each output port along the path through the NoC from the source to the destination tile. In contrast to other NoC types that support *quality of service* (QoS), GS traffic cannot completely block BE traffic, because the WRR scheduling guarantees the selection of a BE time slot in every scheduling cycle, once it has been reserved. A detailed description of the components and the operation principle of the *i*-NoC will be presented in Chap. 4.

2.1.2 *Invasive Language Constructs*

One particular question that has been addressed by language research is how the fundamental programming constructs of invasive computing (*invade*, *infect*, *retreat*) can be embedded into existing programming languages. Instead of extending a language by new keywords and defining their syntax, a library-based approach has been chosen, because it is easier to realize, provides more flexibility, requires less compiler adaptions, and allows ports to different languages. Besides implementations in OpenMP [23] (iOMP [24]) and MPI [25] (iMPI [26, 27]), the main language research has been performed based on the parallel programming language X10 [28], because it provides fundamental language support for concurrency, communication, data distribution as well as synchronization, and follows the PGAS programming model that fits very well to our envisioned heterogeneous tiled architectures. A detailed introduction into X10 including its programming model, language features as well as runtime implementation is provided in Sect. 2.2, since it is the basis for our language (InvadeX10), simulator (InvadeSIM), and actor library (ActorX10) implementation. In [4], Zwinkau introduced InvadeX10, a library-based implementation of the abstract invasive programming constructs in X10, which is presented in the following sections.

2.1.2.1 Invade, Infect, and Retreat

The following program fragment shows the usage of the three basic constructs invade, infect, and retreat as implemented in InvadeX10. It represents a rudimentary invasive program.

```
val claim = Claim.invade(constraints);
claim.infect(ilet);
claim.retreat();
```

The static method `Claim.invade` takes `constraints` and returns an object of the class `Claim`, which gets assigned to the variable `claim` and represents the allocated resources. The class `Claim` provides an `infect` method to distribute and execute computations across the reserved resources. Each call to `infect` requires an *i*-let object, which contains the code to execute together with initial data. The `infect` call blocks the program until all *i*-let computations are finished. A call to the `retreat` method frees all resources from the claim. Afterwards, the claim is empty and further infect calls would not results in any *i*-let execution.

An exemplary definition of an *i*-let is shown in the following code fragment.

```
val ilet = (id:IncarnationID) => {
    Console.OUT.println("Hello! ("+id+")");
};
```

The variable `ilet` gets assigned an X10 function object (also called *anonymous function*), which takes one argument `id`, returns nothing, and prints a greeting to standard output. A function declaration like this is called an *i*-let *candidate*, as it can be used for parallel execution in any infect call, like in the example above. In contrast to C/C++, the X10 language does not provide any function pointers but *closure objects*. Here, any free variable such as `Console.OUT` in the example is bound to its value from the lexical environment at the time of the assignment. Such closure objects are called *i*-let *instances*, since they are bound to a specific resource type. During infection, an *i*-let instance is concurrently copied and distributed to every physical processing resource within the claim. Internally, the `infect` method creates an `IncarnationID` for each processing resource and calls the `ilet` instances each providing its corresponding `id` as argument. Once an *i*-let instance is bound to a concrete physical processing resource, it becomes an *i*-let *incarnation* (that is where `IncarnationID` is derived from). If the operating system scheduler dispatches an *i*-let incarnation for execution on its bound processing resource, it is called *i*-let *execution*. The set of all *i*-let incarnations, which have been created during an `infect` invocation, is called a *team*. Naturally, a team consists of as many *i*-let incarnations as processing resources are available in the corresponding claim.

An essential aspect of resource-aware programming is that applications can express their different demands for resources, which are then used by the runtime system to allocate an appropriate claim for the application. In InvadeX10, the programmer specifies the resource needs by creating and combining different objects of the `Constraint` class. Different subclasses of the `Constraint` class cover various types of constraints [4]. In the following code snippet, the construction and combination of constraints objects is shown.

```
val constraints = new AND();
constraints.add(new PEQuantity(1,8));
constraints.add(new PlaceCoherence());
constraints.add(new MaximumLoad(0.7f));
```

The specified constraints require a number of processing resources between one and eight, all resources have to be located in the same tile, and each resource is only allowed to have a current load of at most 70%. Individual constraint objects can also be combined logically to allow for more specific constraints. For instance, the `AND` constraint acts as a container and is only fulfilled if all added constraints are fulfilled. The `AND` constraint object is assigned to the variable `constraints`, which is later provided as argument to the invade call `Claim.invade`.

2.1.2.2 Invasive Command Space

In order to determine the different flavors of resource-aware programming, more than 25 pseudo-code examples from different domains of parallel as well as embedded computing (e.g., image processing, signal filtering, linear algebra, numerical mathematics, etc.) have been collected and analyzed in [4] to design the command space for the constraint system. Based on this data, the following constraints have been introduced and finally implemented as X10 class hierarchy as depicted in Fig. 2.6.

The first class of constraints is so-called *predicate constraints*, which specify predicates for processing resources. An application might require a processing resource to (1) be under a certain load, (2) be under a certain temperature, (3) have a *floating-point unit* (FPU), (4) have certain amount of local memory, (5) have a scratch pad memory, (6) be of a certain resource type, (7) have a certain cache size, (8) be migratable, or (9) not be scheduled preemptively. Such constraints impose a simple filter operation over the set of available processing resources during an invasion. The second class of constraints is *order constraints*, which provide an order of processing resources according to (1) load, (2) temperature, (3) memory, or (4) speed (in terms of clock frequency). Using these constraints, an application is able to specify a preference, whether it is I/O- or CPU-bound. Specifying multiple of these constraints imposes

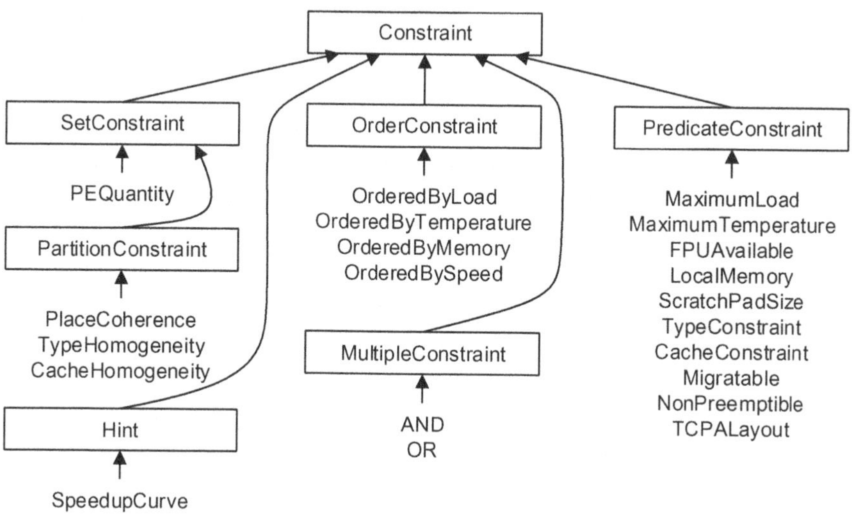

Fig. 2.6 Constraint hierarchy (adapted from [4])

a secondary or ternary order. The third class of constraints is *set constraints*, which specify conditions on a set of processing resources. The most common set constraint is the (1) quantity of processing resources to be claimed. Other set constraints can be summarized as *partition constraints*, which may require (2) place coherence, which means that processing resources are located on the same tile, (3) type homogeneity in terms of the ISA, or (4) cache type homogeneity. Additionally, the two logical operators AND and OR are available to combine constraints. Furthermore, the programmer can specify so-called *hints*, which might contain complex information such as speedup curves of applications and can be used by the runtime system to properly distribute the available resources.

If the specified constraints cannot be fulfilled by the runtime system, a `NotEnoughResources` exception is thrown. A resource-aware application has to handle a situation where an invade call fails due to this exception and has to provide a fallback solution. For an efficient implementation of the AND constraint and an efficient selection of processing resources, the runtime system uses a specific order for evaluating the constraints. Given a set of available processing resources, the following steps are performed:

1. Each resource is filtered according to the predicate constraints.
2. Resources are ordered according to the order constraints in their specified priority.
3. The set of resources is partitioned according to the partition constraints.
4. Select a partition that matches the quantity constraint.

2.1.2.3 Reinvade, Reinfect, and Partial Retreat

Extended concepts of invasive computing are reinvasion, reinfection and partial retreat. Reinfection is to infect a previously infected claim with *i*-lets again. Typically, data is not deleted from the local memory once a team has finished and left the infected processing resources. This is guaranteed, as the runtime system is only allowed to free resources in a retreat phase.

Reinvasion is used, if an application requires additional resources. The resulting claim should unify both sets of resources, the previous as well as the new ones. One option would be to retreat from all resources and call invade with unified constraints. Since this is not an atomic operation, another application might invade resources in the meantime. Thus, InvadeX10 provides unifications of claims: A "delta" claim can be invaded and then merged with the previous claim. An example how an original `claim` can be extended by resources allocated through an additional invasion is shown in the following code snippet.

```
val delta = Claim.invade(constraints);
val claim2 = claim + delta;
claim2.infect(ilet2);
```

The + operator is overloaded to perform the union of the two claims. The subsequent infection simultaneously infects the original as well as the delta claim, while the previously introduced incarnation ids are generated for the unified claim. The original `claim` object is available in addition to `claim2`, so it is still possible to infect just the original set of processing resources.

A partial retreat is used, when the application's DOP is decreased and only a subset of the currently claimed resources is needed for the next computations. In this case, `partialRetreat` can be called at a claim. It requires a constraint object (like in invade) and retreats from resources matching the specified constraints. In the following code example, the amount of processing resources in the `claim` is reduced by four.

```
claim.partialRetreat(new PEQuantity(4));
```

Like in an invade call, a `NotEnoughResources` exception might be thrown, if `claim` contains less than four processing resources.

2.1.2.4 Color Space Transformation Example

The following code example is part of an invasive image compression algorithm chain. The algorithm computes an irreversible color transformation (ICT), which converts an image from the RGB to the $YC_b\,C_r$ color space. Image processing algorithms are candidates for accelerators operating according to the *single instruction, multiple date* (SIMD) principle, like the TCPA as used in invasive computing. Assuming an ICT algorithm specifically tailored to be executed on the TCPA is given by `ComponentTransform.forwardIctTCPA`. The following program shows an example of a resource-aware application trying to invade a TCPA for accelerating an ICT computation, if a TCPA is available on the current tile, otherwise running a sequential ICT computation on the CPU.

```
val img = Image.load(filename);
val constraints = new AND();
constraints.add(new TCPALayout(3,3));
constraints.add(new TypeConstraint(PEType.TCPA));
constraints.add(new PEQuantity(1));
try {
    val claim = Claim.invade(constraints);
    /* Invasion succeeded: Run ICT on TCPA by infect */
    claim.infect((id:IncarnationID) => {
        ComponentTransform.forwardIctTCPA(img);
    });
} catch (e:NotEnoughResources) {
    /* Invasion failed: Run ICT on CPU by local execution */
    ComponentTransform.forwardIctCPU(img);
}
```

Initially, image data is loaded into the tile local memory. The three constraints require a single TCPA providing a rectangle of 3×3 available PEs. The code specific for the TCPA is stored in the `forwardIctTCPA` method. If invasion is successful, the claim containing the TCPA is infected with an *i*-let calling the TCPA specific code given the image data. If no TCPA is available on the current tile, an exception is thrown by the runtime system and has to be caught by a `catch` block as shown in the example. A fallback solution is executed by calling `forwardIctCPU`, which computes the ICT sequentially on the current CPU.

2.1.3 Invasive Run-Time Support System

The introduced concepts of invasion and resource awareness require not only architectural and language support, but also an efficient middleware implementation of these concepts bridging the gap between invasive applications and architectures. Thus, methods, principles, and abstractions for application-controlled invasion on the operating system and run-time level are studied and technically integrated into the *invasive run-time support system* (*i*RTSS). In Fig. 2.7, the different layers related to the *i*RTSS are depicted. The top layer represents invasive applications written in X10 using the InvadeX10 primitives for resource-aware programming. An X10 application internally makes use of the X10 runtime, which has been slightly modified to target the *i*RTSS instead of a standard operating system. Also, the introduced resource-aware primitives of InvadeX10 make use of *i*RTSS functionalities. For example, runtime calls for invading resources, creating new *i*-lets, or sending data to other tiles over the *i*-NoC are re-directed to and handled by the *i*RTSS. The *i*RTSS itself is separated into several layers. The operating system abstraction layer (OSAL) provides the top-level interface to the *i*RTSS used by applications and interacts with the other layers. An agent system is responsible for managing the hardware resources and optimally distributing them to the applications based on their resource requests via `invade`. Agents act on behalf of applications and follow a distributed bargaining protocol to achieve global resource management only based on a local exchange of mapping information. As demonstrated in [10], this approach keeps

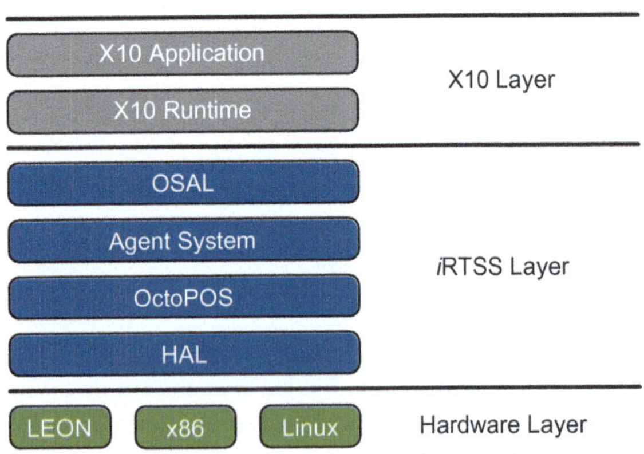

Fig. 2.7 The different layers of the invasive run-time support system

the management overhead almost constant for an increasing number of cores, while centralized approaches do not scale. OctoPOS is a light-weight parallel operating system with an event-based kernel architecture [29]. It provides low-level operating system functionalities such as creating, scheduling, synchronizing, and dispatching *i*-lets supported by the C*i*C and a *remote procedure call* (RPC) mechanism to handle communication over the *i*-NoC. The execution of an *i*-let on a core follows a run-to-completion semantics. Preemption happens only in a cooperative manner if the *i*-let relinquishes processor control due to synchronization with other *i*-lets. A hardware abstraction layer (HAL) is used to provide a generic interface to access the different underlying hardware platforms. Since the *i*RTSS is a very complex software project, but essential to realize the concepts of invasive computing, the full-system simulator InvadeSIM provides an emulation of the *i*RTSS functionality, instead of an exact re-implementation.

2.2 The X10 Programming Language

The X10 programming language is the fundamental basis for our implementations of InvadeX10, InvadeSIM, and ActorX10. Thus, in this section, we provide a detailed introduction to the concepts, language constructs, and runtime implementation of X10 to build the necessary foundation for the following chapters.[4] X10 is an object-oriented, parallel programming language, developed by IBM in cooperation with academic partners [28]. Its development started in 2004 as part of the IBM's *productive easy-to-use reliable computer systems* (PERCS) project which aimed to increase the productivity of parallel application development.

As a consequence to the paradigm-shift from single to multi-core processors, IBM envisioned so-called *non-uniform cluster computing* (NUCC) systems as dominant multi-processor structure in future. Such computing systems are composed of multi-core *symmetric multiprocessing* (SMP) nodes with non-uniform memory hierarchies interconnected by a high-speed cluster network. Our introduced tiled architectures can be seen as miniaturized versions of NUCC systems each integrated onto a single chip. Both types of computing systems pose similar challenges in programming them using conventional methods. Although object-oriented programming languages significantly have improved software productivity by supporting concepts such as portability, type-safety, and automatic memory management, their capabilities in expressing concurrency (e.g., *Java threads*) and distribution (e.g., *Java RMI*) are limited and cannot cope with the requirements for programming NUCC systems. Common practice in the HPC domain is to use a combination of the *message passing interface* (MPI) [25] and *open multi-processing* (OpenMP) [23] to program large distributed memory systems. However, this approach is primarily accessible to system experts rather than application programmers and becomes increasingly inconvenient

[4]The basis for the following syntax and semantics descriptions is the X10 version 2.3.1, which has been also used for all implementations in this book.

and error-prone as software as well as computing system complexity increases. Thus, IBM introduced the object-oriented programming language X10, which by design includes concepts for concurrency and distribution and allows programming NUCC systems efficiently with high productivity.

The key design decisions of X10 were (a) to design a new programming language based on the Java and Scala syntax instead of introducing new libraries or language extensions to gain analyzability and type-safety, (b) to utilize the PGAS programming model for expressing data locality in the form of *places* to address the limited scalability of a single uniform heap, and (c) to introduce dynamic, asynchronous *activities* for expressing concurrency to provide a lightweight and flexible thread alternative. In the following sections, we present an overview of the X10 programming model, the novel sequential as well as parallel language constructs, details about their runtime implementation, and how these concepts are related to invasive computing.

2.2.1 X10 Programming Model

The X10 programming model is based on the *partitioned global address space* (PGAS) model and introduces the fundamental concepts of *places* as a location for computations and data, and *activities* as a lightweight alternative to threads. The PGAS model provides a global address space that is divided into non-overlapping partitions, which are called places in X10. A place is a collection of resident data objects and activities operating on these objects. At run time, each place is mapped to a physical shared-memory location with a certain amount of processing resources, which can be a single computer, a node of a NUCC system, or a tile of a tiled architecture. An activity represents sequential computations that are executed on a place by one of the processing resources. Data objects are called *local* for an activity if they are located on the same place as the activity and *remote* otherwise. In a distributed-memory model, activities may only reference local data, they have no knowledge about remote data, since the address spaces are private to each place. The PGAS model allows activities to reference local as well as remote data like in a global address space of a shared-memory model. However, the places may be physically distributed and a communication between the places is required. An overview of these concepts is depicted in Fig. 2.8. Place-local data resides in the place it was created for its lifetime. Constant data is replicated to and accessible by all places. A special distributed array data structure might be distributed across multiple places and each place holds a local portion of that array.

X10 implements a *globally asynchronous, locally synchronous* (GALS) semantics for reading and writing mutable data. An activity is only allowed to read or write local data objects (locality rule). Any attempt to read or write a remote data object results in a `BadPlaceException`. Activities may only access remote data by spawning an activity at the remote place performing that access. This remote access requires sending an activity from one place to another using a communication mechanism.

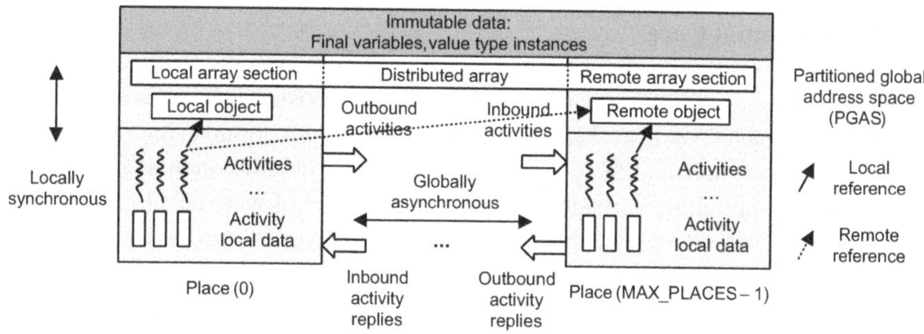

Fig. 2.8 Overview of the X10 programming model (adapted from [28])

Once the access is done, an activity reply is sent back to the initiating place. Due to network latencies, the round-trip time for a remote access might be substantial higher than for a local access. Through the concept of places, X10 makes the difference between fast local and slow remote accesses explicit and allows locality-aware programming. In [28], the authors of X10 state that accesses by activities to local memory locations are guaranteed to be *sequentially consistent*, i.e., read and write operations to a memory location are performed in one indivisible step. Thus, all local write accesses to the same data object are observed in the same order as they have been performed by the activities (locally synchronous). In contrast, remote data accesses have weaker ordering semantics, since sending and receiving of activities happens asynchronously (globally asynchronous). Explicit synchronization primitives have to be used to enforce inter-place *coherence*.

Since X10 allows spawning local as well as remote activities dynamically, it overcomes the limitations of the *single program, multiple data* (SPMD) principle, which is assumed by the PGAS model. Thus, the X10 programming model is an extension to the PGAS model and is also referred to as *asynchronous partitioned global address space* (APGAS). It allows for high-performance and high-productivity programming of heterogeneous computing and memory architectures. Therefore, it fits very well to the needs and the envisioned tiled architectures of invasive computing. Details about the concrete language constructs of X10 and how they are related to the invasive computing primitives are presented in the following sections.

2.2.2 X10 Language Constructs

In this section, important features of the sequential core of X10 as well as concrete X10 language constructs for concurrency, synchronization, and distribution are presented.

2.2.2.1 Sequential Core

The sequential core of X10 is very similar to already existing object-oriented programming languages such as Java, C++, or Scala. As mentioned, the syntax has been adopted from Java and Scala as well as successful features such as portability, type-safety, and automatic garbage collection to increase software development productivity. Besides standard concepts of object-oriented programming such as *classes*, *objects*, *interfaces*, *constructors*, *inheritance*, *overloading* and *overriding* of methods, and basic control flow structures,[5] X10 introduces some substantial extensions to the sequential core, which are extensively used in InvadeX10 and InvadeSIM.

Before explaining the language extensions in detail, the basic declaration of variables and methods in X10 are presented in the following code example, since they are adopted from the Scala syntax and are fundamental to almost every X10 program.

```
val pi = 3.14159; // immutable variable declaration with val
var number : int = 11; // mutable variable declaration with var
def method() {...} // method declaration with def
```

Each variable has to be declared either using `val` or `var`. A `val` variable can be assigned only once and is immutable after assignment. The data type may be omitted if it can be inferred, as explained later in *type inference*. The concrete data object a `val` variable is pointing to might be mutable or immutable. A `var` variable can be assigned multiple times and the data type has to be provided explicitly. In contrast to Java, the order for data type specifications is <identifier>:<data type>. Methods have to be declared using the keyword `def` followed by the method name and parameters. All known modifiers from Java are allowed.

Constrained types X10 Classes may declare some special fields called *properties*, which are immutably bound at the time of instantiation of the class. A type declaration can then formulate certain conditions to these properties, which are checked statically by the compiler. For instance, the following class definition of a matrix (`Matrix`) includes two properties representing the number of rows (`rows`) and columns (`cols`) of the matrix.

```
1 class Matrix(rows:int, cols:int) {
2     operator this * (other:Matrix) {this.cols == other.rows}
          {...}
3 }
```

Furthermore, this class defines a multiplication operator, which allows multiplying two matrices. As known, a multiplication of two matrices requires the number of columns of the first matrix (`this`) being equal to the number of rows of the second matrix (`other`). This can be formulated as Boolean expression using the properties of both matrix objects and added to the operator overloading in curly brackets in line 2. In an X10 program, each call to this operation given two matrices is checked statically and mismatching matrix sizes would cause a

[5] Central control-flow structures of an imperative programming language are `if-then-else` statements, `for` loops, `while` loops, `switch` statements, `try-catch` blocks, and `throw`.

compile error. Thus, it is guaranteed that such constraints are fulfilled already at compile time and dynamic checks are not required.

Generic types Each X10 class may declare *type parameters*. They allow to define generic types, where the abstract type parameters are bound to concrete types at the time of instantiation of the class. Typical examples are container types such as array, list, or heap data structures. The following code snippet shows a section of a simple generic list implementation based on an array.

```
class List[T] {
    val array : Array[T];
    var ptr : int;
    def add(item:T) {
        array(ptr++) = item;
    }
}
```

This list can then be instantiated for different types of contained items, instead of implementing a new list for each item type. Examples are a list of integer numbers `List[int]`, a list of strings `List[String]`, or a list of lists containing floating-point numbers `List[List[float]]`. Many fundamental data structures in InvadeSIM are implemented using generic types.

Function types A powerful concept adopted from functional programming languages are *lambda expressions*, also called *anonymous functions* or *function literals* in X10. They allow to define a block of code like a function with corresponding signature to be used as value. Thus, they can be stored in variables, passed as parameters to functions, or used as an interface definition. The corresponding type of a function literal is called *function type*. Examples of the usage of function literals and types are shown in the following code snippet.

```
1 val square = (n:int) => n*n;
2 val result1 = square(4); // result1 = 16
3 val fun = (f:(int)=>int, n:int) => f(n);
4 val result2 = fun(square, 4); // result2 = 16
```

The function literal stored in `square` calculates and returns the square of a given integer number `n`. The corresponding function type is `(int)=>int`. Line 2 shows how to invoke the function literal for a specific number. Another function literal stored in `fun` requires two arguments: A function `f` of type `(int)=>int` and an integer number `n` of type `int`. It just applies the given function `f` to the given number `n` and returns the result. Line 4 shows how `fun` is applied to the function `square` and 4. InvadeX10 uses function literals for the definition of *i*-lets candidates (see Sect. 2.1.2.1).

Type inference Another concept adopted from functional programming languages is *type inference*. This concept allows omitting an explicit data type specification, e.g., for variable declarations or method return types if the type can be derived by the compiler from the initializer or the method body, respectively. Type inference also derives type parameters of generic types as well as property values of constrained types. Examples of type inference are given in the following code fragment.

```
val n = 42; // n has type int{self==42}
val m = new Matrix(2,3); // m has type Matrix{rows==2, col==3}
val l = new List[float]; // l has type List[float]
def f() { // f has return type boolean
    ... return true; ... return false;
}
```

Type inference gives the most precise X10 type and should be preferred to an explicit definition by the programmer.

Besides the introduced language extensions of the sequential core, X10 also introduces a new type of arrays that are much more complex than arrays from C/C++ or Java. X10 arrays may not only be defined over linear index spaces beginning with 0, but over arbitrary underlying index spaces allowing the efficient creation of sparse and dense multi-dimensional arrays. Furthermore, the data of arrays may be spread across multiple places by using so-called *distributed arrays*. An underlying distribution object determines on which place each element of the distributed array is mapped to. Special statements are provided for an efficient parallel iteration over distributed arrays as well as so-called *collective operations*, which implement common computations on entire arrays such as pointwise operations, reductions, scan, or projection operations.

2.2.2.2 Concurrency

An X10 application it supposed to run on a massively-parallel, distributed-memory architecture and to dynamically create and concurrently execute potentially thousands of parallel threads.[6] However, the conventional concept of *kernel-level* threads such as *portable operating system interface* (POSIX) threads may not provide the required flexibility and performance to efficiently program such parallel systems. Therefore, X10 introduces a lightweight alternative to threads called *asynchronous activities*. In fact, activities are *user-level* threads and much cheaper to create and manage than *kernel-level* threads. Details about the implementation of activities are presented in Sect. 2.2.3.

The X10 language construct to create an activity on the current place is `async {S}`, where `S` denotes the statements that are going to be executed by the activity (the curly brackets might be omitted if `S` represents only a single statement). Activities may be created *locally* or *remotely*, which is explained in Sect. 2.2.2.4. The call returns immediately, since creating an activity is a very cheap operation and the statements `S` are executed concurrently to the calling activity, which continues

[6] A thread is considered as *kernel-level* thread, which is the smallest scheduling unit in an operating system. It is part of an operating system process and might concurrently co-exist beside many other threads in a process. All threads of a process share the same address space and may communicate via shared memory. Kernel-level threads are created and managed by the operating system kernel, which might be very expensive for thousands of threads considering the number of required system calls, context switches, and kernel-level resources.

its execution at the next statement after `async`. Once created, an activity cannot be cancelled by another activity. It is executed until it either terminates regularly or an unhandled exception is thrown. However, it might be blocked during its execution due to synchronizing statements. An activity does not provide any return value, but it may access variables from its surrounding context. In the following code example, the usage of activities is shown.

```
1 def calc(a:float, v:float, t:float) {
2     val t1:float;
3     val t2:float;
4     finish {
5         async t1 = v * t;
6         t2 = (a * t * t) / 2.0f;
7     }
8     return t1 + t2;
9 }
```

A calculation is split into two parts and executed concurrently by two activities. One activity is already being executed, creates another activity in line 5, and immediately continues at line 6. The created activity accesses the variables `t1`, `v`, and `t`, which are defined in its surrounding context. Both activities store their calculated results in the local `val` variables `t1` and `t2`, which can be assigned only once (not necessarily at their declarations). The calling activity waits for the created activity being terminated by using the `finish` construct in line 4, which is explained in Sect. 2.2.2.3. After the created activity has been terminated, the calling activity continues at line 8 and returns the final result. This simple example employs a very fine-grained parallelism, since the activities only execute a couple of operations in parallel. It would not be of practical use, because the overhead of creating and executing the second activity would exceed the gain through parallel execution. However, the concept of activities provides the flexibility to express different granularities of application parallelism such as task, loop, data, or pipeline parallelism.

2.2.2.3 Synchronization

The execution of thousands of concurrent activities may require coordination and synchronization to produce correct results. X10 provides different language constructs for expressing atomicity and certain orderings between activities, which are explained in the following.

`finish {S}` is the first ordering construct and realizes a *barrier* synchronization. It has been already used in the code example from Sect. 2.2.2.2. An activity may wait for the termination of all recursively created activities by putting a `finish` block around the statements `S` creating new activities. The barrier applies to locally as well as remotely created activities. If uncaught exceptions are thrown by the child activities, they are aggregated at a `finish` block and thrown after the barrier has been left (rooted exception model). The code example from Sect. 2.2.2.2 is repeated here to highlight the `finish` usage.

```
def calc(a:float, v:float, t:float) {
    val t1:float;
    val t2:float;
    finish {
        async t1 = v * t;
        t2 = (a * t * t) / 2.0f;
    }
    return t1 + t2;
}
```

If multiple activities concurrently operate on shared data structures, *mutual exclusion* is required to preserve data integrity and to avoid *race conditions*. For this purpose, X10 provides the `atomic {S}` construct, which realizes place-local *atomic sections*. If an activity executes the statements `S` protected by an `atomic` block, they are executed atomically. During an atomic access, no other activity on the same place is allowed to enter the atomic section. They might be suspended and wait until the atomic section has been left by the accessing activity. Conceptually, atomic blocks are executed in a serialized order with respect to all other atomic blocks on the same place (weak atomicity). The following code fragment shows an extract of a linked list implementation.

```
1 class LinkedList {
2     var head : Node;
3     def add(data:int) {
4         val node = new Node(data);
5         atomic {
6             node.next = head;
7             head = node;
8         }
9     }
10 }
```

This code allows adding new elements to the front of a linked list concurrently. Creating a new `Node` is not protected, since it operates on activity-local data. However, the access to the `head` pointer is protected by an `atomic` block in line 5, since it represents shared data. The `head` pointer is saved in the new `node` object and is then overridden by it. Without this atomic block, elements could get lost in case of concurrent accesses. The statements `S` of an `atomic` block must follow certain rules, otherwise an exception is thrown. They must not execute blocking operations (e.g., `finish`), must not create further activities (e.g., `async`), and must not access remote data (e.g., `at`). In fact, they are only allowed to execute sequential control flow.

In addition to the presented *unconditional* atomic block, X10 also provides a *conditional* atomic block by introducing the `when (c) {S}` construct. This construct allows an activity to wait until a certain condition `c` becomes true and then to execute the statements `S` atomically. The unconditional `atomic {S}` block can be seen as the special case `when (true) {S}`. For the statements `S` of a `when` block, the same restrictions hold as for an `atomic` block. The following code example represents a single-element *first in, first out* (FIFO) buffer that may be used to synchronize the data exchange between concurrently running activities.

```
1 class Buffer {
2     var data : int = 0;
3     var filled : boolean = false;
4     def push(n:int) {
5         when (!filled) {
6             data = n;
7             filled = true;
8         }
9     }
10    def pop() {
11        val n:int;
12        when (filled) {
13            n = data;
14            data = 0;
15            filled = false;
16        }
17        return n;
18    }
19 }
```

Activities may produce or consume a single integer data element by calling `push` or `pop` at the FIFO buffer. The `when` construct easily allows realizing blocking behavior for the operations if the buffer is either full or empty. If `push` is called, the state of the buffer is checked by `when` in line 5. If `filled` is `false`, the statements in line 6 and 7 are executed immediately (data is stored in the buffer and `filled` is set to `true`). If `filled` is `true`, the current activity is blocked until another activity consumes the data element by calling `pop`. The activity is unblocked once the other activity has left the `when` block from line 12. The condition is re-evaluated and the statements in line 6 and 7 are executed. The `pop` method behaves analogously to the `push` method. The state of the buffer is checked by `when` in line 12. If `filled` is `true`, the statements from line 13 to 17 are executed immediately (the data element is stored temporarily, `data` is set to `0`, `filled` is set to `false`, and the data element is returned). If `filled` is `false`, the current activity is blocked until another activity produces a data element by calling `push`.

An advanced feature for coordinating activities are `Clocks`. They introduce a general barrier concept that is much more flexible than the `finish` construct. A `finish` block waits for the termination of all enclosed activities. Clocks allow to split up a parallel computation into coordinated phases. Each phase is finalized by a barrier synchronization between all participating activities (on possibly different places) once they have finished their computations in this phase. However, the activities are not terminated, instead they may exchange partial results and continue their computations in further phases. In contrast to the `finish` construct, activities may be dynamically added to or removed from a group of activities synchronized by a clock. Furthermore, an activity may simultaneously participate in several clock-synchronized computations.

2.2.2.4 Distribution

X10 programs target complex, distributed-memory architectures, which requires proper notions of concurrency and distribution to efficiently program such systems. Based on the PGAS programming model, X10 introduces the fundamental concept of *places* to describe the distribution of data and computations. A place is a collection of data objects and activities operating on these objects and is mapped to a physical shared-memory location at run time. Details about the implementation of places are presented in Sect. 2.2.3. Data is called *local* for an activity if it is located on the same place as the activity, otherwise it is called *remote*. In contrast to a distributed-memory model, the PGAS model allows activities creating, referencing, and accessing not only local, but also remote data objects (remote operations).

X10 provides the *place-shift* construct `at (p) {S}` to fundamentally support remote operations. This construct synchronously executes the statements `S` on place `p`. Internally, the current activity sends a message to place `p` and waits until a reply is received. On place `p`, an activity is created and executes the statements `S`. Once the activity is terminated, it sends a reply back to the initial place, which unblocks the waiting activity. The `at` construct may be executed asynchronously by using `at (p) async {S}`, which does not wait for the statements `S` being finished, but only creates a remote activity and immediately continues locally. The syntax of the `at` construct is very compact and easy to understand, but the semantics is quite complex. Thus, multiple code examples are provided to highlight different aspects of the semantics and present common use cases of the `at` construct.

Since we consider heterogeneous architectures, some places may provide more computing power than others. The `at` construct may be used to outsource expensive computations to a more powerful place to speed them up. In the following example, the calculation from Sect. 2.2.2.2 is now executed remotely instead of spawning another activity on the same place.

```
1 def calc(a:float, v:float, t:float) { // calc is called on Place(0)
2     val result = at (Place(1)) {
3         v * t + (a * t * t) / 2.0f
4     };
5     return result;
6 }
```

The method `calc` is assumed to be called on `Place(0)`. The class `Place` is the representation of an X10 place on language level to identify and address them. The set of available places is linearly ordered and indexing starts with 0. Thus, `Place(0)` is the initial place, where an X10 application starts (also `Place.FIRST_PLACE`). In line 2, a place shift to `Place(1)` is applied and the calculations in line 3 are executed on `Place(1)`. The result is sent back to `Place(0)`, stored in `result`, and finally returned in line 5. Important to note is that the parameters `a`, `v`, and `t` are copied to `Place(1)`, since they are accessed by the statement in line 3.

As for an `async` call, the statements `S` of an `at` call may access any variable in its surrounding context. However, in contrast to `async`, each accessed variable is copied to the target place (distributed object model) and may lead to unexpected behavior as presented in the following example.

```
1 val array = new Array[int](1); // array is created on Place(0)
2 array(0) = 47;
3 at (Place(1)) {
4     val t1 = array(0); // t1 = 47
5     val t2 = t1 - 36; // t2 = 11
6     array(0) = t2;
7 }
8 val t3 = array(0); // t3 = 47
```

In line 1, an array of integer numbers (`Array[int]`) with a size of 1 is created on `Place(0)` and stored in the variable `array`. The first element of the array is set to 47 in line 2. After switching to `Place(1)` in line 3, the first array element is read and stored in `t1`. As expected, `t1` contains the number 47, since the array has been copied to `Place(1)`. Then, a simple calculation is applied (line 5) and the first array element is changed to 11 (line 6) on `Place(1)`. After returning to `Place(0)`, the first array element is unexpectedly still 47, although it should be 11 as changed in line 6. However, due to the copy semantics, two different arrays exist (one on `Place(0)` and one on `Place(1)`). They do not represent the same data objects, even though they are accessed by the same name and the second array is initialized with the data from the first array. Modifying the content of the array on `Place(1)` does not change the content of the array on `Place(0)`. Furthermore, data is copied in X10 according to the *deep copy* semantics, i.e., not only the accessed variables are copied to the target place, but also the entire object graph that are recursively referenced by these variables.

In order to reference and access a unique data object created on a certain place from any place, a so-called *global reference* (`GlobalRef[T]`) has to be used. A global reference is a pointer to a concrete data object created on a specific place. It may be freely copied across different places and will always point to the unique data object it was created from. The following code snippet shows the example from above using global references.

```
// array is created on Place(0)
val array = GlobalRef[Array[int]](new Array[int](1));
array()(0) = 47;
at (Place(1)) {
    val t1 = at (array.home) array()(0); // t1 = 47
    val t2 = t1 - 36; // t2 = 11
    at (array.home) array()(0) = t2;
}
val t3 = array()(0); // t3 = 11
```

Again, an array of integer numbers with a size of 1 is created on `Place(0)`. This array is wrapped by a global reference (`GlobalRef[Array[int]]`), which is stored in the variable `array`. In order to access the actual array data, the global reference has to be de-referenced by calling the *parenthesis operator* such as `array()`. This operator can only be called successfully on the place on which the wrapped data object has been created (the so-called *home* place), otherwise a

`BadPlaceException` is thrown. This home place is immutably stored as `home` property in each global reference at the time of its creation. In line 3, the first element of the array is set to 47 using the parenthesis operator. This time, switching to `Place(1)` in line 4 does not result in copying the entire array to `Place(1)`, but only its global reference. To access the unique array data on `Place(0)` from `Place(1)`, a place shift to its home place is required and applied in line 5 to copy the first array element to `Place(1)`. At the home place of the array (`Place(0)`), the global reference is de-referenced and the first array element is read. It is returned by the `at` call and stored in `t1`, which contains the number 47, as expected. A simple calculation is applied on `Place(1)` (line 6) and the result is copied to `Place(0)` by the `at` call (line 7) to change the first array element to 11. Finally, after the control has been returned to `Place(0)` and the first array element is read in line 9 and it correctly contains the number 11.

Using the `at` construct and global references allow creating, referencing, and accessing remote data and to spread complex computations to multiple places. The number of places available to an X10 program is specified before the program is started and the X10 runtime is responsible for the mapping of places to physical locations in a NUCC system (nodes) or in a tiled architecture (tiles). On X10 language level, the number of available places is accessible via `Place.MAX_PLACES` and the set of places may be obtained by `Place.places()`. Furthermore, X10 provides convenient constructs to simplify the distribution of data such as *distributed* as well as *remote* arrays.

2.2.3 X10 Runtime

As presented in the previous sections, X10 introduces the APGAS programming model, activities, places, and a variety of novel language constructs. In order to implement the presented concepts, X10 uses a combination of compiler techniques and runtime support. The X10 runtime system is the bridge between the X10 programming language and the underlying parallel architecture. In order to target the type of tiled architectures considered in invasive computing and simulated by InvadeSIM, a couple of modifications to the X10 runtime system have been applied as part of the *i*RTSS emulation in InvadeSIM. Thus, in this section, we first explain the standard implementation of activities and places by the X10 runtime and then how the presented behavior may be modified to target a simulated invasive architecture.

As explained already, activities are lightweight alternatives to threads and much cheaper to create and manage. In the standard X10 runtime, they are executed by so-called X10 *worker threads*. These are kernel-level threads of the underlying operating system (e.g., POSIX threads) and scheduled by the operating system scheduler to the available cores of the target machine. Each worker thread is equipped with an individual activity queue and processes this queue by permanently popping activities and executing them. If a new activity is created, it is pushed into the activity queue of the currently executing worker thread. After an activity has been popped, it is

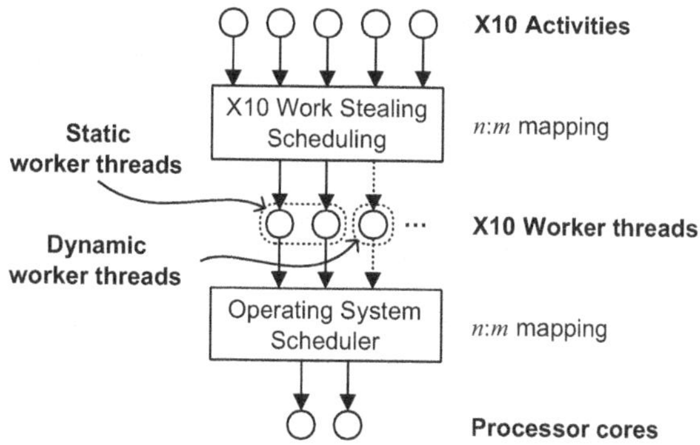

Fig. 2.9 Scheduling of activities on a single place by several worker threads

executed until it terminates before the next activity is processed (run-to-completion semantics). If an activity queue is empty, the corresponding worker thread tries to steal activities from other worker threads' activity queues. This scheduling approach is called *work stealing* and improves the workload balance among the worker threads. A general overview of the activity scheduling on a single place is depicted in Fig. 2.9. The advantage of this scheduling approach is that expensive kernel-level threads are created only once and then process their activity queues. No contexts have to be switched between two activity executions of one worker thread and the creation of an activity is just a local queue operation. This efficiently allows creating thousands of activities being executed by only a couple of worker threads. The idea is to enable scalable and adaptable X10 programs that can be easily mapped without modifications onto different target architectures with a different amount of processor cores. Two types of worker threads exist: *static* and *dynamic*. Static worker threads are created at the beginning of the program and exist until it terminates. Their number is fixed at run time and is specified via the environment variable `X10_NTHREADS`. A general practice is to create as much static worker threads as processor cores are existing. A dynamic worker thread is created on demand at run time if an activity has executed a blocking operation (e.g., `finish` or `when`). This ensures always a free worker thread that may unblock the waiting activity through the execution of other activities. After unblocking, this worker thread is not required anymore and deactivated. However, it is kept for later uses and becomes a *spare* thread.

In the standard X10 runtime, each place is mapped to an operating system process. An operating system process has its own address space and provides a shared-memory location for the activities running on that place. Furthermore, each process has its own worker threads for activity execution. The number of places available to an X10 program may be specified via the environment variable `X10_NPLACES`. The X10 runtime system allows mapping all places onto a single target machine as depicted in Fig. 2.10a or onto multiple machines as depicted in Fig. 2.10b. If all places are mapped to one machine, each place gets its own operating system process and all worker threads from all processes compete for the available processor cores. Communication between the places is realized by machine-local *inter-process communication* (IPC).

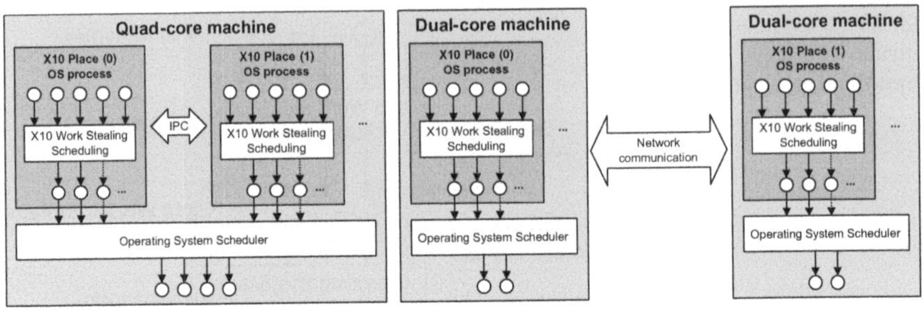

(a) Single quad-core machine **(b)** Two dual-core machines

Fig. 2.10 Mapping of places to different target machines configurations

In case of a multi-node target machine, a configuration file specifies which place is mapped to which machine each identified by an internet protocol (IP) address. The processes on the machines are created accordingly. Worker threads compete for processor cores only within their local machine and communication between places is realized by a network communication mechanism between machines and by IPC within a machine. More details about the X10 communication mechanism (X10RT) is presented in Chap. 4.

Although X10 introduces new parallel language constructs, the main implementation of them is not realized by the compiler, but by the runtime system. The X10 compiler is a source-to-source translator, which transforms an X10 program depending on the used back end into a C++ or a Java program, which is then compiled by a native compiler to the respective target machine.[7] One pass of the X10 compiler replaces each parallel language construct by a call to the X10 runtime system. This approach allows modifying the behavior of the parallel language constructs easily since the X10 runtime system is represented as X10 class (`x10.lang.Runtime`) and can be adapted and extended easily. A compiler modification would be much more difficult. For instance, the submission of a created activity (`Runtime.runAsync()`) may be modified to target another scheduler than the standard X10 scheduler. An overview how the X10 compiler replaces each parallel language construct to X10 runtime calls is provided in Table 2.1. They represent essential locations, which have to be adapted in order to target different runtime implementations such as the *i*RTSS.

2.2.4 Relation to Invasive Computing

After the concepts, language constructs, and runtime implementation of X10 have been presented in detail, the relation to the concepts of invasive computing is briefly summarized in this section. First, X10 introduces the fundamental concepts of places

[7] All X10 implementations and modifications presented in this book are based on the C++ back end.

Table 2.1 Mapping of parallel X10 language constructs to X10 runtime calls

X10 language construct	X10 runtime call
`async {S}`	`Runtime.runAsync({S});`

`finish {S}`	`Runtime.startFinish();`
	`{S}`
	`Runtime.stopFinish();`

`atomic {S}`	`Runtime.enterAtomic();`
	`{S}`
	`Runtime.leaveAtomic();`

`when (c) {S}`	`Runtime.enterAtomic();`
	`while (!c)`
	`Runtime.awaitAtomic();`
	`{S}`
	`Runtime.leaveAtomic();`

`at (p) {S}`	`Runtime.runAt(p, {S});`
`at (p) async {S}`	`Runtime.runAsync(p, {S});`

as part of the PGAS programming model to represent data and computations. In invasive computing, the tiles of a tiled architecture represent the physical mapping targets for places, since they contain a certain amount of processing resources for executing activities as well as the TLM2 as location for place-local data and shared-memory communication. In comparison, the mapping targets of places in a NUCC system are the nodes of the system. The second fundamental concept introduced by X10 are activities to represent concurrency. Invasive computing introduces *i*-lets as parallel execution entities. In fact, an *i*-let candidate is represented by an X10 anonymous function and just represents a parallel code section, but no execution. The `infect` call is responsible for the distribution of *i*-lets to the corresponding processing resources in the claim. During this procedure, an *i*-let candidate becomes an *i*-let incarnation after it has been mapped to a concrete processing resource. An *i*-let incarnation may be realized by an activity. However, activities need to be extended as follows to be used as *i*-let incarnations. They need to have a claim *identification* (ID) to control their executions on tiles and it must be possible to pin them to specific processing resources. Both extensions have been done in InvadeSIM. An activity is then executed by a particular processing resource and calls the anonymous function representing the *i*-let functionality. In comparison, the execution of activities in a NUCC system is realized by worker threads. Finally, any communication between places via `at` is realized in invasive computing by sending messages between the corresponding tiles over the *i*-NoC. In a NUCC system, messages are sent between nodes using a certain network communication library.

2.3 Summary

In this chapter, we introduced the novel computing paradigm invasive computing, which has been proposed to tackle current and upcoming challenges in programming and designing highly-parallel many-core architectures. The main idea is to allow applications to map and distribute their workload themselves onto a many-core architecture based on their temporal computing demands as well as the temporal availability and other state information of the underlying hardware resources (e.g., power, temperature, or faultiness). We introduced the considered type of tiled architectures including concepts of invasion on architecture level, the realization on language level through InvadeX10, and the different software layers of invasive run-time support system to support invasive computing concepts on the operating-system level. These different implementation levels have to be properly modeled in InvadeSIM to allow full-system simulation of invasive computing systems. Furthermore, we introduced the programming model, language constructs, runtime implementation of the parallel programming language X10. The introduced concepts for expressing concurrency through activities and distribution through places serve as the fundamental implementation basis to InvadeX10, InvadeSIM, and ActorX10.

References

1. J. Teich, Invasive algorithms and architectures. It Inf. Technol. **50**(5), 300–310 (2008). ISSN: 1611-2776. https://doi.org/10.1524/itit.2008.0499
2. J. Teich et al., Invasive computing: an overview, in *Multiprocessor System-on-Chip: Hardware Design and Tool Integration*, ed. by M. Hübner, J. Becker. (Springer, Heidelberg, 2011), pp. 241–268 (Chap. 11). ISBN: 978-1-4419-6459-5. https://doi.org/10.1007/978-1-4419-6460-1
3. J. Teich et al., Invasive computing-concepts and overheads, in *Proceeding of the Forum on Specification and Design Languages (FDL), Vienna, Austria, 18–20 Sept 2012* (IEEE, 2012), pp. 217–224. ISBN: 978-1-4673-1240-0
4. F. Hannig et al., Resource-aware programming and simulation of MPSoC architectures through extension of X10, in *Proceedings of the 14th International Workshop on Software and Compilers for Embedded Systems (SCOPES), St. Goar, Germany, 27–28 June 2011* (ACM, 2011), pp. 48–55. ISBN: 978-1-4503-0763-5. https://doi.org/10.1145/1988932.1988941
5. A. Tanase et al., Exploitation of quality/throughput tradeoffs in image processing through invasive computing, in *Proceedings of the International Conference on Parallel Computing (ParCo), Munich, Germany, 10–13 Sept 2013* (IOS Press, 2013), pp. 53–62. https://doi.org/10.3233/978-1-61499-381-0-53
6. S. Roloff et al., Invasive computing for predictable stream processing: a simulation-based case study, in *Proceedings of the 13th IEEE Symposium on Embedded Systems for Real-time Multimedia (ESTIMedia), Amsterdam, The Netherlands, 8–9 Oct 2015* (IEEE, 2015). ISBN: 978-1-4673- 8164-2. https://doi.org/10.1109/ESTIMedia.2015.7351761
7. S. Wildermann et al., Invasive computing for timing-predictable stream processing on MPSoCs. It Inf. Technol. (IT) **58**(6), 267–280 (2016). ISSN: 1611-2776. https://doi.org/10.1515/itit-2016-0021

8. J. Turek, J.L. Wolf, P.S. Yu, Approximate algorithms scheduling parallelizable tasks, in *Proceedings of the 4th Annual ACM Symposium on Parallel Algorithms and Architectures (SPAA), San Diego, CA, USA, June 29–July 1 1992* (ACM, 1992), pp. 323–332. ISBN: 0-89791-483-X. https://doi.org/10.1145/140901.141909
9. D.G. Feitelson et al., Theory and practice in parallel job scheduling, in *Job Scheduling Strategies for Parallel Processing (JSSPP)*. ed. by D.G. Feitelson, L. Rudolph. (Springer, Heidelberg, 1997), pp. 1–34. ISBN: 978-3-540-69599-8. https://doi.org/10.1007/3-540-63574-2_14
10. S. Kobbe et al., DistRM: distributed resource management for onchip many-core systems, in *Proceedings of the 9th IEEE/ACM/IFIP International Conference on Hardware/Software Codesign and System Synthesis (CODES+ISSS), Taipei, Taiwan, 9–14 Oct 2011* (IEEE, 2011), pp. 119–128. ISBN: 978-1-4503-0715-4. https://doi.org/10.1145/2039370.2039392
11. S. Wildermann, T. Ziermann, J. Teich, Game-theoretic analysis of decentralized core allocation schemes on many-core systems, in *Proceedings of the Conference on Design, Automation and Test in Europe (DATE), Grenoble, France, 18–22 Mar 2013* (EDA Consortium, 2013), pp. 1498–1503. ISBN: 978-1-4503-2153-2
12. A. Tanase et al., On-demand fault-tolerant loop processing on massively parallel processor arrays, in *Proceedings of the 26th IEEE International Conference on Application-specific Systems, Architectures and Processors (ASAP), Toronto, Canada, 27–29 July 2015* (IEEE, 2015), pp. 194–201. ISBN: 978-1-4799-1925-3. https://doi.org/10.1109/ASAP.2015.7245734
13. V. Lari et al., Providing fault tolerance through invasive computing. It Inf. Technol. **58**(6), 309–328 (2016). ISSN: 1611-2776. https://doi.org/10.1515/itit-2016-0022
14. V. Lari et al., Hierarchical power management for adaptive tightly-coupled processor arrays. ACM Trans. Des. Autom. Electron. Syst. (TODAES) **18**(1), 2:1–2:25 (2013). ISSN: 1084-4309. https://doi.org/10.1145/2390191.2390193
15. H. Khdr et al., Power density-aware resource management for heterogeneous tiled multicores. IEEE Trans. Comput. (TC) **66**(3), pp. 488–501 (2017). ISSN: 0018-9340. https://doi.org/10.1109/TC.2016.2595560
16. H.-J. Bungartz et al., Invasive computing in HPC with X10, in *Proceedings of the 3rd ACM SIGPLAN X10 Workshop (X10), Seattle, WA, USA, 20 June 2013* (ACM, 2013), pp. 12–19. ISBN: 978-1-4503-2157-0. https://doi.org/10.1145/2481268.2481274
17. M. Bader, H.-J. Bungartz, M. Schreiber, Invasive computing on high performance shared memory systems, in *Facing the Multicore-Challenge III: Aspects of New Paradigms and Technologies in Parallel Computing*, ed. by R. Keller, D. Kramer, J.-P. Weiss. (Springer, Heidelberg, 2013), pp. 1–12. ISBN: 978-3-642-35893-7. https://doi.org/10.1007/978-3-642-35893-7_1
18. J. Heisswolf et al., The invasive network on chip-a multi-objective many-core communication infrastructure, in *Workshop Proceedings of the 27th International Conference on Architecture of Computing Systems (ARCS), Lübeck, Germany, 25–28 Feb 2014* (VDE Verlag, 2014), pp. 1–8. ISBN: 978-3-8007-3579-2
19. J. Henkel et al., Invasive manycore architectures, in *Proceedings of the 17th Asia and South Pacific Design Automation Conference (ASP-DAC), Sydney, NSW, Australia, Jan 30–Feb 2 2012* (IEEE, 2012), pp. 193–200. ISBN: 978-1-4673-0770-3. https://doi.org/10.1109/ASPDAC.2012.6164943
20. J. Henkel et al., i-Core: a run-time adaptive processor for embedded multi-core systems, in *Proceedings of the International Conference on Engineering of Reconfigurable Systems and Algorithms (ERSA), Las Vegas, NV, USA, 18–21 July 2011* (2011), pp. 1–8
21. F. Hannig et al., Invasive tightly-coupled processor arrays: a domain-specific architecture/compiler co-design approach. ACM Trans. Embed. Comput. Syst. (TECS) **13**(4), pp. 133:1–133:29 (2014). ISSN: 1539-9087. https://doi.org/10.1145/2584660
22. J. Heisswolf, R. König, J. Becker, A scalable NoC router design providing QoS support using weighted round robin scheduling, in *IEEE 10th International Symposium on Parallel and Distributed Processing with Applications (ISPA), Leganes, Spain, 10–13 July 2012* (IEEE, 2012), pp. 625–632. ISBN: 978-1-4673-1631-6. https://doi.org/10.1109/ISPA.2012.93
23. L. Dagum, R. Menon, OpenMP: An industry standard API for shared-memory programming. IEEE Comput. Sci. Eng. **5**(1), 46–55 (1998). ISSN: 1070-9924. https://doi.org/10.1109/99.660313

24. M. Gerndt et al., Invasive computing with iOMP, in *Proceeding of the Forum on Specification and Design Languages (FDL), Vienna, Austria, 18–20 Sept 2012* (IEEE, 2012), pp. 225–231. ISBN: 978-1-4673-1240-0
25. Message Passing Interface Forum. *MPI: a message-passing interface standard*. Technical Report (1994)
26. I.A. Comprés, M. Riepen, M. Konow, M. Gerndt, Invasive MPI on intel's single-chip cloud computer, in *Proceedings of the 25th International Conference on Architecture of Computing Systems (ARCS), Prague, Czech Republic*. ed. by A. Herkersdorf, K. Römer, U. Brinkschulte. (Springer, Heidelberg, Feb 28–Mar 2 2012), pp. 74–85. ISBN: 978-3-642-28292-8. https://doi.org/10.1007/978-3-642-28293-5_7
27. I.A. Comprés et al., Infrastructure and API extensions for elastic execution of MPI applications, in *Proceedings of the 23rd European MPI Users' Group Meeting (EuroMPI), Edinburgh, United Kingdom, 25–28 Sept 2016* (ACM, 2016), pp. 82–97. ISBN: 978-1-4503-4234-6. https://doi.org/10.1145/2966884.2966917
28. P. Charles et al., X10: an object-oriented approach to non-uniform cluster computing, in *Proceedings of the 20th Annual ACM SIGPLAN Conference on Object-oriented Programming, Systems, Languages, and Applications (OOPSLA), San Diego, CA, USA, 16–20 Oct 2005* (ACM, 2005), pp. 519–538. ISBN: 1-59593-031-0. https://doi.org/10.1145/1094811.1094852
29. B. Oechslein et al., OctoPOS: a parallel operating system for invasive computing, in *Proceedings of the International Workshop on Systems for Future Multi-Core Architectures (SFMA), Salzburg, Austria, 10 Apr 2011* (2011), pp. 9–14

Chapter 3
InvadeSIM-A Simulation Framework for Invasive Parallel Programs and Architectures

In the previous chapters, we highlighted the trend towards NoC-based heterogeneous MPSoC designs implementing hundreds to thousands of processing resources on a single chip, including standard RISC processors, accelerators as well as specific processors with extended instruction sets. To cope with upcoming challenges in designing, programming, and managing such architectures, *Invasive Computing* has been introduced, a novel paradigm for resource-aware computing, which proposes self-organization of parallel applications on an MPSoC by using concepts for requesting, executing on, and releasing processing as well as communication resources. These concepts have shown to utilize the underlying resources more efficiently by exploiting time-varying degrees of application parallelism. In order to study the dynamic behavior of invasive programs and to investigate a broad range of different invasive architectures without the need for a full hardware and software implementation, fast and flexible simulation techniques are mandatory.

Therefore, one main contribution of this book is the design and development of the full-system simulator InvadeSIM, which allows modeling complex invasive architectures, emulating the execution behavior of the invasive runtime system, and simulating function and timing of invasive parallel applications apart from utilization, efficiency, and competition. Important research questions that have to be answered in order to realize such a full-system simulator are: What is the right level of abstraction to model processors, accelerators, memory accesses as well as network communication to allow simulating complex application scenarios, while preserving not only functional correctness, but also other aspects such as accuracy and how combining these components into a unified simulation environment?

A well-established method in the area of platform design for MPSoC is virtual prototyping. Here, virtualized software development environments are provided, which allow for flexible design and high-speed simulation of processor-based systems. However, these tools are mainly targeted to support early software development before real hardware is available and the effort to build and update the underlying

S. Roloff et al., *Modeling and Simulation of Invasive Applications and Architectures*, Computer Architecture and Design Methodologies,
https://doi.org/10.1007/978-981-13-8387-8_3

models is quite high. Also, the integration of custom components such as dedicated hardware accelerators into such frameworks is difficult. Other approaches for implementing full-system simulators utilize *transaction-level modeling* (TLM) [1, 2] within SystemC or even cycle-accurate models. Such simulators typically operate on the *electronic system level* (ESL) [3] and provide very detailed timing information. However, they also require very detailed models of the simulated hardware, often do not support heterogeneous multi-core architectures, and finally are much too slow for our purposes. In the area of single processor simulation, state-of-the-art approaches employ *instruction set simulation*, *binary code translation*, or *compiled simulation*. They achieve very high simulation speeds on the functional level, but when accurate timing information is required, the simulation speed might be slowed down dramatically. Here, *trace-driven simulation* [4] or *sampling techniques* [5, 6] can be applied. They are widely used in system-level design and support efficient design space exploration. However, such approaches are often restricted to homogeneous shared-memory multi-core architectures, and there exist hardly any simulator that is able to simulate heterogeneous processor architectures with invasive capabilities.

Therefore, in this chapter, we present a novel high-level, high-performance simulation framework called InvadeSIM, which tackles the complexity and the heterogeneity of future MPSoCs while enabling the investigation of the programming paradigm invasive computing. Proposed is a timing-approximate processor simulation approach, which is based on a discrete-event, direct-execution simulation and a timing estimation mechanism called *time warping* to scale the elapsed execution times from the host processors to the simulated target processors. Furthermore, timing models for accelerators such as TCPAs as well as a contention model of concurrent accesses to the tile-local shared memory are presented. The integration into InvadeSIM including the scheduling of *i*-lets to the processors is also addressed. Besides NoC simulation, which will be presented in the subsequent chapter, these concepts for simulating a single tile represent one fundamental building block of InvadeSIM. An architecture description file is used to individually configure large heterogeneous tiled architectures. It allows describing the number and arrangement of the tiles as well as the individual tile components such as processors and accelerators including clock frequencies, hardware costs, and other parameters. The basis of the contributions of this chapter has been published in a couple of peer-reviewed papers. An early prototype of a functional simulation has been presented in [7]. The time-warping mechanism and the synchronization approach have been introduced in [8] and extended by a memory-congestion model in [9]. Before explaining the details of InvadeSIM, we provide a brief classification of general processor simulator organizations.

3.1 Simulator Organizations

To evaluate next-generation computer architectures, system designers heavily rely on execution-driven full-system simulators that strive for functional fidelity that allows executing relevant workloads and performance fidelity that enables to compare and rank different design alternatives. However, the trend toward more complex system architectures including deeper processor pipelines, higher number of integrated cores, caches, and memory, or even network-based communication structures, make the design and implementation of new system simulators increasingly difficult. Another challenge for execution-driven simulators is the massive exploitation of thread-level parallelism of modern computer architectures, which can be realized by multiprocessor, multi-core, or hardware multithreading designs. In such systems, there exist multiple correct program executions depending on the interleaving of the threads. However, the order in which the threads are executed or interleaved significantly influences the program performance and highly depends on the system's timing. For instance, assuming two concurrent invasive applications competing for exclusive resource accesses on an invasive architecture. Depending on the implementation of the resource management, the decision whether a resource is assigned to the first or the second application is made internally in a critical code section, which is typically protected by a lock. The application that acquires the lock first will also receive the resource and creates a new thread, will probably have a different performance or even executes another algorithm as without that resource. The other application is blocked until the lock is free again and has to stick to the remaining resources. Therefore, in order to accurately model these timing-dependent outcomes in a modern multi-core simulator, it is essential to precisely determine the actual thread interleavings. Furthermore, in order to achieve high functional fidelity, it is not sufficient to allow only single-threaded applications to be simulated. Instead, simulators should include runtime or operating system functionality to simulate several competing multi-threaded applications on a heterogeneous MPSoC. To manage the increasing simulator complexity, the authors of [10] propose different decoupled simulator organizations that separate functional and performance concerns, which are depicted in Fig. 3.1. The goal of such decoupled organizations is to reduce complexity and thereby increase flexibility and productivity. However, they may reduce simulator performance due to the interactions between the separate components. They are briefly explained in the following according to the definitions from [10].

Integrated simulator couples the models for timing and functionality in one monolithic unit. It is required to provide functional implementations combined with respective timing information for all system components. Such simulators can be very detailed by modeling several stages of a processor pipeline or cache effects, and due to this tight coupling, timing-dependent thread interactions are well captured. However, they lack flexibility as new devices or performance models have to be integrated. Potential bugs are difficult to isolate and to fix. Its complexity and inflexibility reduce productivity and make it difficult to balance accuracy and performance.

Functional-first simulator generates in a first step a logical sequence of instructions that have been produced by functional simulation. In a second step, these instructions are fed into a separate timing simulation. Well-known examples of this simulation technique are trace-driven and static full-system simulators. Here, traces of events, either instructions or system calls to the operating system, are first

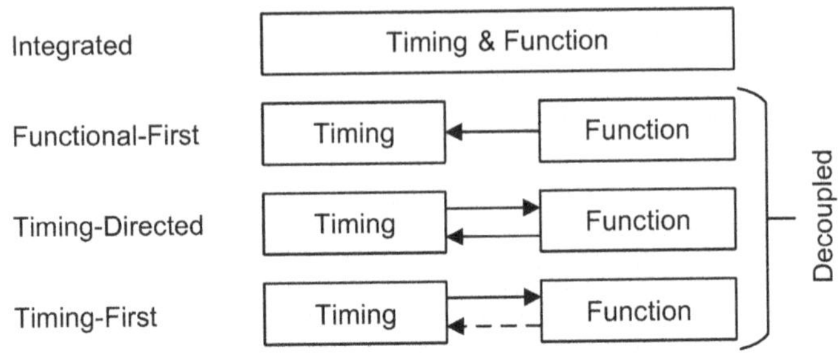

Fig. 3.1 Execution-driven simulator organizations (adapted from [10])

recorded and then replayed in another component. This decoupled organization reduces the complexity of the individual components, but only allows simulating a single execution path and makes it impossible to consider timing-dependent thread interactions.

Timing-directed simulator lets a timing simulator direct a functional simulator. This organization allows executing speculative paths and considering thread interleavings correctly. It comes at the costs of a more complex functional simulator that has to support program execution at multiple alternative paths. The coupling of timing and function in this design is tighter than in other designs, since the timing simulator can choreograph partial functional execution. It has similarities with the integrated simulator, but is still decoupled.

Timing-first simulator lets a timing simulator run ahead of a functional simulator. It is similar to the timing-directed organization, but here, the timing simulator is more complex and has its own simplified functional implementation. It determines the timing of the operations, manipulates its own simplified state, and verifies its functional execution by consulting the functional simulator from time to time. If a deviation is detected, the timing simulator state has to be repaired. This approach improves simulator performance through a reduction of detailed functional simulator calls. It is also able to model speculative execution and timing-dependent thread interactions.

Our novel simulation approach does not fit into only one of the presented categories. Rather, it is a combination of functional-first and timing-directed simulator and will be explained in detail in the following sections.

3.2 Direct-Execution Simulation

Enabling to study the complex interactions of multiple programs concurrently executing on an invasive computing architecture or to compare the performance of different architecture alternatives or different claim assignment strategies requires a simulation environment that is driven by the execution of real workloads,[1] rather than by statistical workload models or traces derived from the sequential execution of the programs. Here, it must be easy to change and customize the heterogeneous architecture description in terms of, e.g., tile number, tile organization, number of cores per tile, etc. Data-dependent computation times, interprocess communications, and synchronization delays shall be accounted for a high degree of accuracy. However, simulating such complex systems including heterogeneous architectures with hundreds of cores, the emulation of an invasive runtime system, and the execution of realistic workloads requires a significant amount of computation time and memory. These requirements on full-system simulation and real workload execution prohibit the usage of instruction set or even cycle-accurate simulation techniques. Instead, we propose a simulation approach based on *direct execution*, which has its roots in the Rice Parallel Processing Testbed [11], the Wisconsin Wind Tunnel I [12] and II [13], and the BigSim simulator [14]. In direct execution, a parallel program for the target system is executed directly on the simulation host machine. Timing information about the program's execution is dynamically generated and calculated by the host. Direct execution can run orders of magnitude faster than pure software simulators, where every target instruction is interpreted and provides a trade-off between flexibility, simulation performance close-to-real-time and sufficient timing accuracy for architecture evaluations.

In Fig. 3.2, the difference between instruction set and direct execution simulation is illustrated. Figure 3.2a shows the basic cycle of an instruction driven simulation. Here, the simulator loads a binary or an assembly program for the target architecture as input and executes a separate cycle for each simulated processor. One cycle simulates the activity on a single processor as long as instructions are available. First, a target processor instruction including command and operands is fetched from the instruction stream and if necessary decoded. The behavior of this instruction is now emulated by modifying the target processor state, which accounts for much of the large overhead associated with this type of simulation, since multiple host instructions must be executed to simulate a single target instruction. Last, the target time of the processor is advanced by the time required to execute that instruction. In case of a multi-core target architecture, the instruction streams from the different target cores are simulated in a round-robin manner and changed after each instruction.

Figure 3.2b shows the cycle of the direct execution simulation approach. Here, the execution of a program (functional simulation) and the simulation model for the architecture (timing simulation) are interleaved. First, a process for simulation is selected and sequentially executed on the host machine until an interaction is encountered. This interaction may be communication, synchronization, or another event manipulating the simulation state. The time required for the last sequence of instructions is determined by the host and added to the target time of the processor. Furthermore, the functional and timing simulation of the processor interaction is performed. Also, an event queue insertion has to be performed and the next process

[1]Considered as realistic workloads are invasive X10 applications that dynamically request, release, and occupy computation and communication resources using InvadeX10.

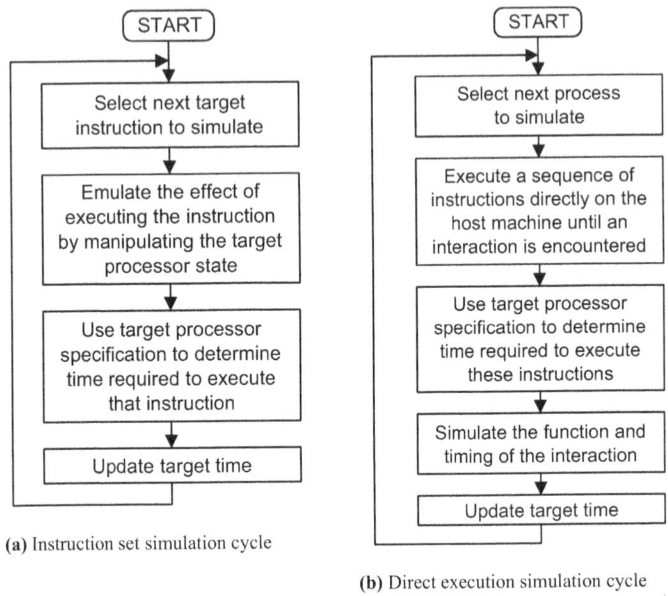

Fig. 3.2 Flow diagrams of the cycles for instruction set simulation and direct-execution

to be simulated is selected. A major advantage of this simulation approach is that there is almost no overhead required to simulate the activity within a processor. Only the interactions between the processors need to be simulated in detail. The cycle for direct execution simulation is significantly faster than the corresponding cycle for instruction driven simulation of parallel systems, because it executes the instructions of the simulated processors instead of simulating their operations and it requires much less event queue operations.

The authors of the Wisconsin Wind Tunnel II [13] identified four key operations that a discrete-event, direct-execution simulator has to address:

1. Calculation of the target execution time,
2. Simulation of features of interest,
3. Communication of target messages, and
4. Synchronization of simulation processes.

As one of the main contributions of this book, all four key-operations have been successfully realized and finally been integrated into a simulator called InvadeSIM. In the following sections of this chapter, the realization of the key-operations 1, 2, and 4 will be explained. Besides timing models for general-purpose processors, key-operation 1 also includes timing models for invasive-specific processors such as *i*-Cores and TCPAs as well as timing models for local memory contention and dynamic frequency adaption. Key-operation 2 includes the basic tile-local invasive runtime functionality such as creating new activities or *i*-lets, starting and stopping them, and activity synchronization via atomic sections. Communication-related tile-crossing runtime calls and an efficient NoC simulation model is the essence of key-operation 3 and will be presented in Chap. 4. A synchronization framework for the ordered execution of different simulation processes such as hardware simulation (NoC), software simulation (processors), or special tasks (e.g., for the scheduling of pre-defined external events at a certain point in time) will be presented in this chapter as the realization of key-operation 4. Finally, the parallelization of this synchronization framework will be presented in Chap. 5.

3.3 Calculation of the Target Execution Time

Simulation is generally uninteresting without any notion of the target's execution time. In contrast to emulation, simulation should not only provide a functional model but also an estimation of the performance of a physical target system. For instruction-set simulation, calculating the target execution time is simple, because the simulator only has to update a clock variable after each simulated target instruction. However, this does not hold for direct-execution simulation, because blocks of sequential host instructions have to be transformed into an estimation of the execution time on the target machine. Besides detailed cycle-accurate simulation, the authors of BigSim [14] propose three different approaches to estimate the execution time of sequential code blocks. They are listed in ascending order with respect to accuracy and complexity:

1. User-supplied expressions,
2. Wall clock measurements, and
3. Hardware performance counters.

In the first approach, a code block is annotated by an expression, which is evaluated at run time and returns an estimation of the execution time for this block on the target machine based on the given parameters. It is very simple, yet flexible. The second approach measures the wall clock time on the host machine that is required for the given code sequence. It can be further multiplied by a suitable scaling factor to obtain a predicted total execution time. The last approach uses hardware performance counters on the host machine to determine a better approximation of the execution time than the previous approaches. Different hardware events such as floating point, integer, or branch instructions can be counted and mapped via a heuristic to an estimation of the total computation time. We used the last approach for a general-purpose processor timing simulation and the first approach for a TCPA timing estimation.

3.3.1 Time Warping–Timing Estimation for General-Purpose Processors

In this section, we present a performance estimation technique for general-purpose processors, which allows predicting the timing of sequential code blocks of previously calibrated target processors based on their execution on the host machine.

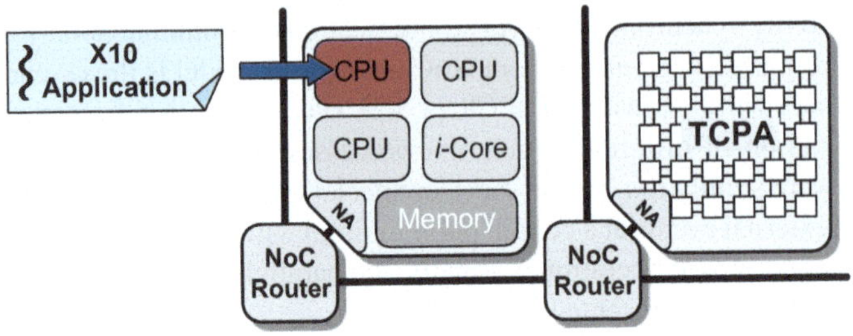

Fig. 3.3 Simulation of a single-activity X10 application on one target core of an invasive multi-tile MPSoC architecture

Considered are single-activity X10 applications, which are composed of only one activity, multi-activity X10 applications are addressed in Sect. 3.3.2. Our timing model is a heuristic approach and does not explicitly model the internal structure of a processor core or consider a particular type of an ISA to derive timing information. Instead, it is based on performance counter measurements and uses a set of pre-determined weights to map the performance counter values to an execution time on the target processor. It can model any target processor including LEON cores as used in invasive computing (see Fig. 3.3), but also more complex ones such as an Intel Core i7 and delivers important timing information based on the real execution of applications. It is very flexible, fast, and sufficiently accurate for our purposes of analyzing the behavior of invasive applications and architectural design decisions.

In Fig. 3.4, the general idea and flow of our direct-execution simulation approach including processor performance estimation are illustrated. The timing model only considers the sequential parts of an X10 application without interruption. Higher-level concepts of X10 and invasive computing for parallelization, synchronization, communication or reservation are handled on higher simulation levels and will be explained later. First, an X10 application is compiled on the host machine into an executable, which also includes an *invasive runtime system*. After starting the binary on the host and loading the target architecture from a template file, the simulated invasive runtime system determines a claim, respectively an appropriate target core for the application (invade) and virtually starts it there (infect). Effectively, the application is executed on the host machine, and hardware performance counters collect information during its execution. Once an interaction has been encountered, or the application has finished, they stop collecting information and create a footprint of the counted events for this executed code section. This footprint is fed into a linear performance estimation model specific to the simulated target core to predict the execution time for the executed code section on that core. The target time of this core is advanced by the predicted time value, the encountered event is processed, and the next code section will be simulated. This procedure is repeated until the simulation of the application has finished. Similar approaches based on performance counters have already been successfully applied in the context of predicting the run-time

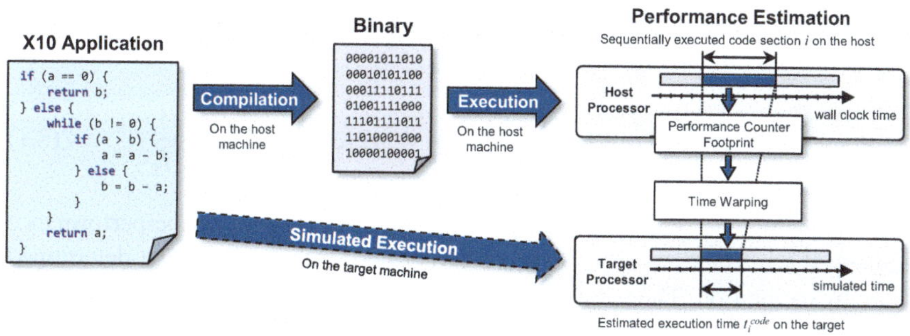

Fig. 3.4 Direct-execution simulation approach including processor performance estimation

power consumption of CPUs [15–19]. In the following, we explain our proposed performance counter-based timing prediction model in more detail.

Performance counters are present in almost every modern (host) CPU as part of the *performance monitoring unit* (PMU). They are architectural registers that allow programmers to gain low-level insights into system performance. They can be configured via programming libraries (e.g., *perfmon2* [20]) or kernel-level interfaces (e.g., *perf_event* [21]) to count the occurrence of certain micro-architectural events during the execution of a program. Effectively, they neither introduce any timing overhead nor interfere with the application's execution at run time. Commonly supported events include the number of committed instructions, clock cycles, cache misses, or branch mispredictions. However, not all of the events can be used to derive a timing model, because most (host) platforms only allow counting a maximum of four events in parallel. Therefore, we propose a 3-predictor linear performance estimation model based on the following three performance counter events:

- c_0: Retired instructions,[2]
- c_1: L1 data cache read misses, and
- c_2: L1 data cache write misses.

They primarily represent characteristics of the application, rather than of the host hardware details and have a major correlation to the application's execution time. Although, the used host machine-compiler pair including compile flags (e.g., optimizations) has a non-negligible influence on the generated instruction stream. Thus, a calibration for each target core and host machine-compiler pair is proposed. To reduce operating system-related timing overhead on the host machine at run time, the kernel, idle, and hypervisor modes are excluded from counting. Important to note is that these performance counter events are not influenced by dynamic frequency scaling on the host machine. Assuming a linear correlation between these performance counter values and the application's timing behavior, we propose the

[2]Retired instructions are the actual number of instructions that have been executed for a particular program flow, i.e., without speculative executed instructions.

following expression to determine the predicted execution time $t_{i,j}^{code}$ for the sequentially executed code section i on target core j:

$$t_{i,j}^{code} = \left(p_{j,0}\left(c_{i,0} - \left(c_{i,1} + c_{i,2}\right)\right) + p_{j,1}c_{i,1} + p_{j,2}c_{i,2}\right)\frac{1}{f_j} \tag{3.1}$$

Here, the parameters $p_{j,0}$, $p_{j,1}$, and $p_{j,2}$ are constant weights for the performance counter values. They are specific to the target core j and have to be determined offline by calibration. The parameter f_j is the current simulated clock frequency of target core j and can be adapted at run time, which will be discussed in detail in Sect. 3.3.4. During the execution of an application, the simulated target core might be changed according to scheduling decisions of the invasive runtime system. Thus, we define the mapped target core of code section i by $\beta(i) \in [0, m-1]$, where m is the total number of cores. Assuming, the application is composed of n sequential code sections implying $n-1$ processed interactions, the estimated total execution time $\hat{T}$ of the application can be expressed as follows:

$$\hat{T} = \sum_{i=0}^{n-1} t_{i,\beta(i)}^{code} + \sum_{i=0}^{n-2} t_i^{interaction} \tag{3.2}$$

Here, $t_i^{interaction}$ is the simulated time the event of code section i requires for being processed. This might be blocking times through atomic accesses, sending messages, or user-defined sleep times. In this case, the application is removed from execution and will be re-scheduled after the event has been processed. Again, the invasive runtime system decides on which target core it will continue its execution. This simulation approach allows modeling large heterogeneous architectures easily by providing different calibrated weights for each type of target core. The timing behavior of invasive applications running on such architectures turn out to be sufficiently accurate and can be quickly determined. In the following section, we exemplary model an Intel Core i7 and evaluate our timing estimation approach according to simulation accuracy and performance by running a benchmark of twelve X10 applications on a single target core.

Example: Timing Calibration for the Intel Core i7

In this section, we show how to determine the weights p_0, p_1, and p_2 of our timing model for an Intel Core i7 processor. We simulate example applications from an X10 benchmark on this calibrated target core and compare the predicted execution times with the sequential execution times measured on a real Intel Core i7 machine. We also evaluate the slowdown S that is introduced by the simulation compared to native execution. The slowdown is an important performance metric for simulators, it is defined by the ratio of the required wall clock execution time T' of the simulator on the host machine to simulate the application to the execution time T of the application on the target machine:

$$S = \frac{T'}{T}$$

Table 3.1 Calibrated performance counter weights for the Intel Core i7 processor

Parameter	Value
p_0	0.39
p_1	8.17
p_2	3.89

Calibrating a new target core requires a reference implementation of that core, either by a real machine, a FPGA implementation, or even a cycle-accurate simulator. For both systems, the host machine and the reference implementation of the target core, a set of X10-based microbenchmarks is compiled. Those microbenchmarks are small X10 programs that have only a single sequential code section and do not require any interaction. A nominal frequency on both systems has to be selected and dynamic frequency scaling has to be deactivated. Each microbenchmark is then executed on the host and on the target machine and a tuple of the three performance counter values c_0, c_1, and c_2 from the host machine and the measured execution time T on the target machine is created. These samples are then used to determine the best fitting parameters p_0, p_1, and p_2 of Eq. (3.1) by minimizing the *mean squared error* MSE through linear regression. The resulting parameters for the Intel Core i7 processor can be found in Table 3.1.

After calibrating the host machine-target core pair, we evaluate the accuracy and performance of our simulation approach by comparing the native execution with the simulation results of a real-world benchmark of twelve X10 applications. These applications are far more complex than the microbenchmarks and are composed of hundreds of sequential code sections and interactions. For each benchmark program, we determined the measured execution time T on the target machine as well as the estimated execution time $\hat{T}$ using our simulation approach on the host machine. The results can be seen in Fig. 3.5. In total, an average error of 18% can be observed for these applications according to the following error calculation:

$$E = \frac{|\hat{T} - T|}{T}$$

Furthermore, the determined slowdown for each benchmark program is also depicted in Fig. 3.5. Here, it can be seen that the maximum slowdown is less than 5 and the average value is 1.76. For some programs, the determined slowdown is smaller than 1. This can happen because event processing on the simulated runtime system might be a bit faster than on the real runtime system. Summarizing, these results show that our presented simulation approach is accurate enough for analyzing the sequential timing behavior of real-world applications by only requiring moderately more wall clock time for simulation than for native execution. In the next sections, we will present, how parallel applications on several cores on a single tile are simulated.

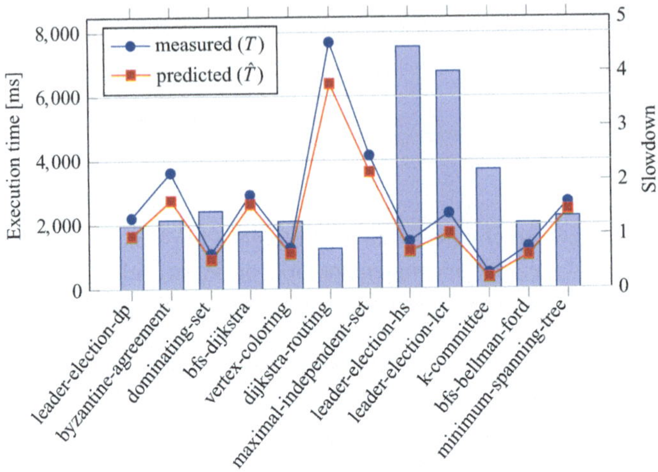

Fig. 3.5 Results of the measured and predicted execution times and the resulting slowdown of our proposed direct-execution and timing-estimation simulation approach for a benchmark of twelve complex X10 applications

3.3.2 Synchronization of Simulation Processes

In the previous section, we discussed the timing simulation of a single-activity X10 application, no synchronization with other activities has been considered. In this section, we introduce a generic DES framework, which allows synchronizing multiple simulation processes, which are also called *simulation tasks* in the following. This includes the simulation of multi-activity X10 applications on a single tile as shown in Fig. 3.6, but even on multiple tiles. Besides the simulation of an activity on a core by direct-execution, a simulation task might also be the simulation of a message transmission in a NoC, an accelerator timing simulation, or the scheduling of an external event at a certain point in time.

Like in each DES, synchronization between several simulation processes is required to obtain correct simulation results. Moreover, all events of all simulation tasks have to be processed in the right timing order. Otherwise, *causality errors*[3] may happen. Missing timing synchronization could also lead to wrong functional results. For instance, processing of resource reservation requests in the wrong order can lead to totally different resource assignments and finally to unexpected simulated application behavior. Though, violating the causality constraint does not necessarily result in simulation errors, e.g., if the events are independent of each other. However, analyzing the dependencies between all dynamically generated events is almost

[3]Causality describes the relationship between cause and consequence for a sequence of associated events. If event B is induced by event A, then, A is the cause of the consequence B and emerged temporally earlier than B. The inversion of this principle is called causality error. An example might be if the reply of a communication request arrives earlier than the request was sent.

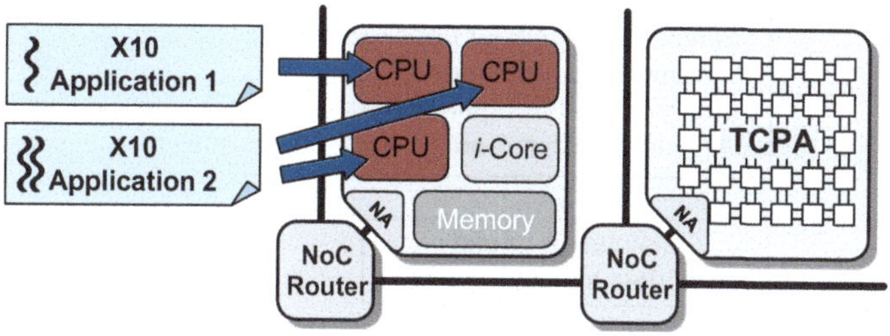

Fig. 3.6 Simulation of a single-activity and a multi-activity X10 application on a total of three target cores

impossible. This is why we use a global timing ordering of all events in our DES framework.

In order to achieve global timing synchronization, all simulation tasks are maintained and scheduled by a simulation scheduler. Each simulation task manages its own local simulated time. A min-heap data structure is used to sort the simulation tasks by their local times. It allows removing, adding, and re-ordering them in $\log n$ time complexity, which is essential if hundreds of target cores are envisioned. The scheduler selects the task with the smallest timestamp by removing the head from the heap and assigns it to a free simulation thread for execution. A detailed flow chart of our synchronization approach is depicted in Fig. 3.7. Simulation threads basically act as workers for executing simulation tasks. One simulation thread can only execute a single simulation task at the same time. In this section, only a single simulation thread is considered, which results in a sequential DES. Our proposed parallel simulation concepts, which utilize several simulation threads to exploit also modern multi-core host machines, are presented in Chap. 5. A task execution might result either in hardware or software simulation or just in an event execution, which allows simulating asynchronous events such as faults or exceptions at a particular point in time. A task execution phase ends with a so-called *synchronization point*, which is initiated by the task itself. The task enters a synchronization phase, where its local simulated time is advanced according to the simulation timing result and timing synchronization with the other simulation tasks is requested. It is again inserted into the heap and the control is returned to the simulation scheduler, which selects the next task for execution. Each simulation task alternates between these two phases of execution and synchronization until it is finally finished. In case of a cycle-accurate hardware simulation, the execution of an event would result in a manipulation of the hardware state and the local time would be advanced by the time of a single cycle.

In case of a software simulation task, our direct-execution simulation approach is applied by the functional simulation of the corresponding code section until an interaction is encountered followed by the timing estimation to advance the local time and the synchronization with the other simulation tasks before the interaction can be processed. In comparison, instruction-set simulation of several cores synchronizes after each instruction to switch between the cores. Our direct-execution

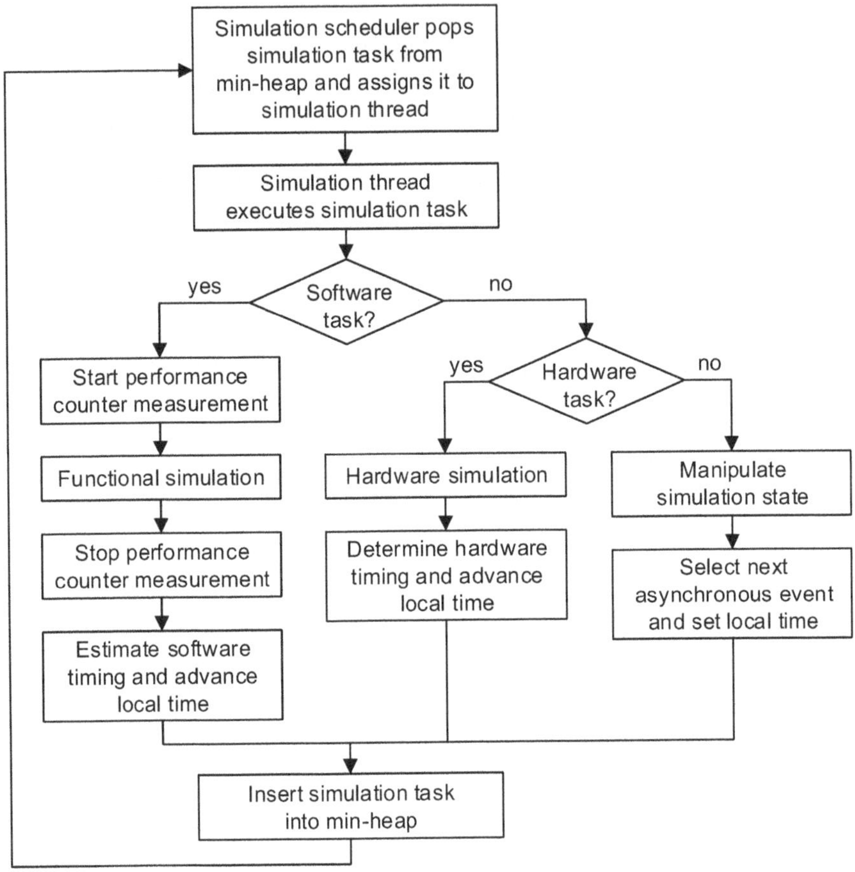

Fig. 3.7 Overview of the synchronization flow in our DES framework

simulation approach only synchronizes at application-dependent interactions, which occur thousands of times less frequent than each instruction. Examples of such interactions are:

- Creating or terminating an activity on the current tile,
- Sending messages to other tiles,
- Entering and leaving atomic sections via locks,
- Calling system functions for returning the current simulated target system time or for sending an activity to sleep, and
- Low-level calls to the invasive runtime system such as changing the frequency of a core, controlling the execution of the TCPA or the *i*-Core.

High-level calls to the invasive runtime system such as invading, infecting, or releasing resources are realized by a software layer, which is also simulated and makes

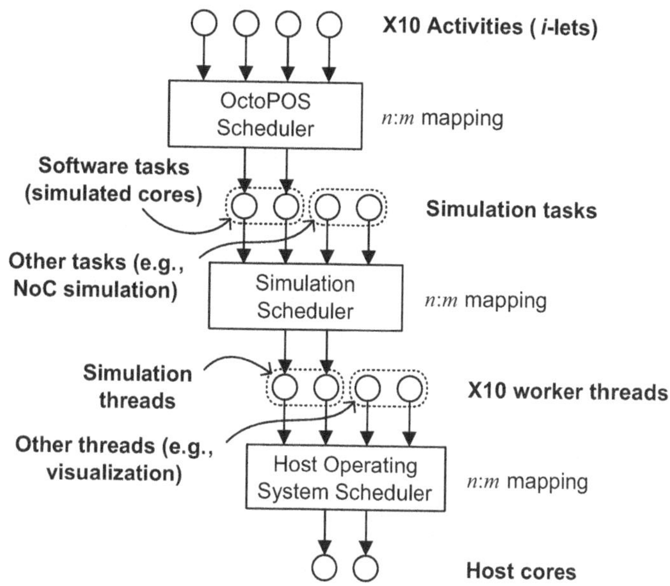

Fig. 3.8 Overview of the scheduling hierarchy in InvadeSIM

use of the introduced low-level functions. Hence, such high-level calls might result in hundreds of interactions.

In order to clarify the relation between X10 activities (*i*-lets), simulation tasks, simulation threads, and X10 worker threads, Fig. 3.8 gives an overview of the scheduling hierarchy realized in InvadeSIM. Concerning software simulation, the highest scheduling level is the simulated operating system scheduler (OctoPOS). It manages all tile-local activities and schedules them to the available target cores on the corresponding tile. Each target core of a tile is simulated by one software simulation task, which are globally synchronized and scheduled by the simulation scheduler of our DES framework, which represents the middle scheduling level. At this level, also hardware simulation and asynchronous event processing are handled. The scheduling targets are simulation threads, which in fact are X10 worker threads finally executing simulation tasks. X10 worker threads are scheduled by the host operating system scheduler to the available host cores in arbitrary orders. It represents the lowest scheduling level. These threads have no relation to the simulated time and can be executed almost independently of each other. Correct timing synchronization must be guaranteed at the middle level. X10 worker threads are also used for purposes other than simulation, e.g., for periodic, non-timing related functions such as visualization or garbage collection. They are not simulated and run in parallel to the simulation threads, but they can access the simulation state, which is for instance required by the visualization of the target hardware state.

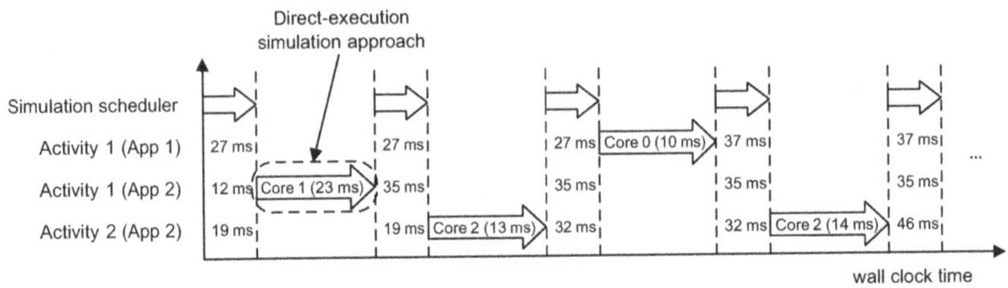

Fig. 3.9 Example simulation and synchronization of three activities on one tile as depicted in Fig. 3.6 by a single simulation thread

Example: Sequential Simulation of Three RISC Cores on One Tile

In Fig. 3.9, an exemplary visualization of the simulation and synchronization of the three activities from Fig. 3.6 is shown. Each activity is mapped to another RISC core on the depicted tile. The diagram shows the execution flow of one simulation thread, which alternately switches between the scheduler and simulation task execution. The local simulated time of each activity is shown in its corresponding row in the scheduler column. Initially, the local simulated times of the three activities are 27, 12, and 19 ms, as shown in the first scheduler column. In this case, activity 1 from application 2 has the smallest timestamp and is selected by the scheduler for simulation. The simulation thread switches the context to this activity, which is indicated by the dashed line, and starts the simulation via direct-execution on core 1. The simulation is interrupted by a synchronization point and the timing estimation for this code section resulted in 23 ms, which is added to the local time of this activity. The context is switched to the simulation scheduler and now activity 2 from application 2 is selected and the simulation process is continued. It can be seen that the simulation of activities is advanced stepwise at application-specific synchronization points and synchronized globally, which includes any processor on any tile. In this scenario, only a single simulation thread was used, the parallel simulation will be discussed in Chap. 5.

3.3.3 *Shared Memory Bandwidth Contention Model*

In the previous section, we introduced a generic DES framework for the synchronization and simulation of several activities on different cores as well as tiles. However, the introduced processor timing model does not consider the interference between different cores accessing a shared memory such as highlighted in Fig. 3.10. Each memory controller provides a certain bandwidth to serve the memory accesses by the cores. If one or more activities from the same or different applications extensively access the shared memory, then, they might overutilize the available memory bandwidth and are slowed down, because of the contention at this shared resource. Thus,

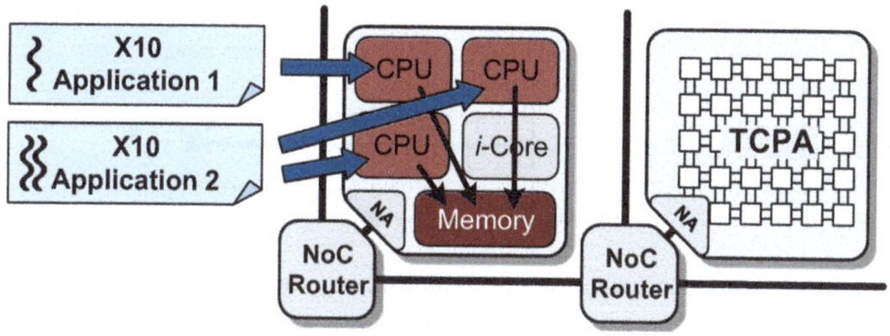

Fig. 3.10 Simulation of three target cores being delayed by contention effects due to concurrent accesses to the tile-local shared memory

in this section, we introduce a simple shared memory bandwidth contention model[4] to simulate this slowdown effect, if the available memory bandwidth is exceeded by the shared memory accesses of the cores.

The actual accesses to the shared memory highly depend on the cache hierarchy of the architecture and the memory access pattern of the application. This is why we present a heuristic approach similar to the timing model as presented in Sect. 3.3.1, which uses performance counter footprints to derive an approximation of the required memory bandwidth by each accessing core and the resulting slowdown. It does not model a detailed cache hierarchy, but delivers important timing information for the usage in architecture simulation. Here, it is especially important to evaluate performance effects of different parallelization concepts to determine an optimal balance between local memory accesses and remote NoC communication.

To illustrate the idea of our memory bandwidth contention model, Fig. 3.11 shows a snapshot of the simulation of three cores accessing the shared memory on a tile. The black dots represent synchronization points and the black lines the resulting processor latencies for the code sections on the target cores according to our direct-execution simulation approach. The code sections on core 0, 1, and 2 have been simulated from their initial until their next synchronization point. In addition to the processor timing result, we determine an estimation for the required memory bandwidth for each simulated code section on the cores. In our example, this is 3 GB/s (core 0), 5 GB/s (core 1), and 12 GB/s (core 2). According to our synchronization approach, core 0 was simulated latest, because its initial synchronization point was bigger than that of the other cores. At this simulation state, only the marked interval between t_0 and t_1 is considered for the memory bandwidth calculation. Congestion delays from earlier code sections are assumed to be already considered until t_0. The upper bound t_1 is the time of the current synchronization point of core 1, because it is the earliest next synchronization point and the simulation will continue from there. Within this interval, the estimated memory bandwidths are known for all three cores and are summed up to a total required bandwidth of 20 GB/s. The simulated memory controller exemplary provides a bandwidth of 16 GB/s. Thus, the resulting average slowdown for this interval can be calculated by dividing the required by the available

[4]Assuming a round-robin bus arbitration (no prioritization).

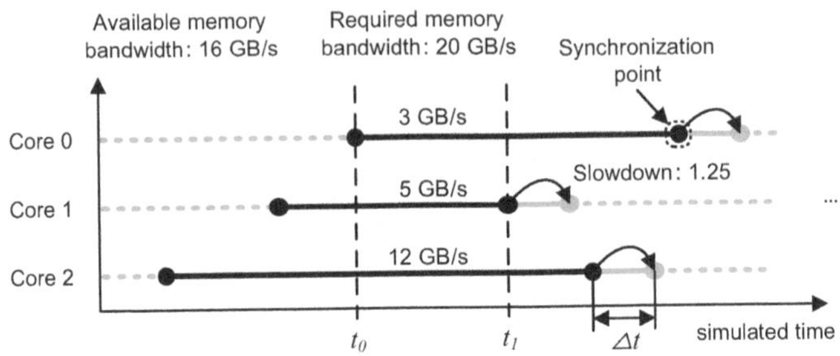

Fig. 3.11 Illustration of our shared memory contention model by an exemplary simulation of three cores having different individual bandwidth requirements for their simulated code sections

memory bandwidth. If this ratio is larger than 1, the cores have to be delayed by a small time increment Δt. In our example, the average slowdown for all cores is 1.25. The local simulated times of the corresponding simulation tasks as well as their position in the min-heap are updated accordingly. The simulation is continued with the updated timestamps and the procedure is repeated for the next code section.

In the following, we explain, how we calculate an estimation of the required memory bandwidth for a simulated code section using the performance counter footprint as already introduced in Sect. 3.3.1. Similar to Eq. (3.1) for the execution time estimation $t_{i,j}$, we propose the following expression to determine an estimation of the required memory bandwidth $b_{i,j}$ for the code section i on the target core j:

$$b_{i,j} = \frac{(q_{j,1} \cdot c_{i,1} + q_{j,2} \cdot c_{i,2})}{t_{i,j}} \tag{3.3}$$

The parameters $c_{i,1}$ and $c_{i,2}$ are the L1 data cache read and write misses, determined by the performance counter measurement. $q_{j,1}$ and $q_{j,2}$ are constant weights for the performance counter values and have to be calibrated for each target core type. The calculated memory bandwidth is an average value for the last code section and does not take into account an irregular distribution of the memory accesses. A description of the memory contention computation is given in Algorithm 3.1. This algorithm is applied after a software simulation has been interrupted by a synchronization point and the execution time $t_{i,j}$ and the memory bandwidth $b_{i,j}$ has been calculated for the current code section i on the current target core j. It determines the total required memory bandwidth of all active (non-idle) cores on the current tile and calculates the slowdown given the available tile memory bandwidth. In case of a memory bandwidth saturation, the execution times of all active cores are increased by the time increment Δt and their positions in the min-heap are updated.

Algorithm 3.1 Computes the shared memory bandwidth contention on a tile after a software simulation task has entered a synchronization point.

```
function ApplyBandwidthContention(tile, t0)
  cores = getActiveCores(tile)
  t1 = min_{∀core∈cores} {getTime(core)}
  if t1 − t0 > 0 then
    slowdown = Σ_{∀core∈cores} getBandwidth(core) / getBandwidth(tile)
    if slowdown > 1 then
      Δt = (t1 − t0) (slowdown − 1)
      for all core ∈ cores do
        addTime(core, Δt)
         updateMinHeap(core)
```

3.3.4 Dynamic Frequency Adaption

In today's computing systems, *dynamic voltage and frequency scaling* (DVFS) techniques are widely used for *dynamic power management* (DPM). It allows adapting the clock frequency or the voltage of processor cores at run time to reduce the overall power consumption or adapting to the current needs of the applications. In invasive computing, applications may have phases of different computing requirements. For example, in one phase, an application may need only very little computing power and fast execution times are not important. Then, frequencies of the invaded cores can be reduced to save power. However, in another phase, the application may have to finish a task very quickly and requires a lot of computing power. Then, the frequencies of the invaded cores can be increased to their maximum level to provide the fastest possible execution time, at the price of high power consumption.

Our current timing estimation for the execution time of processor cores according to Eq. (3.1) includes the clock frequency of the core. This parameter affects the timing of the respective simulated core and hence, the cores competing for shared memory bandwidth. Changing it asynchronously at simulation time either by a simulated hardware component or by a software interface requires special treatment due to the principles of our simulation approach. Thus, in this section, we present a method to adapt the clock frequency of target cores at run time while maintaining correct timing and synchronization of all affected cores. An illustration of the idea is depicted in Fig. 3.12. Three cores are simulated each running at an initial clock frequency of 25, 66, and 120 MHz. By using a software interface, the activity running on core 1 increases the frequency of core 0 from 25 to 50 MHz at t_0, decreases the frequency of core 2 from 120 to 80 MHz at t_1, and changes its own frequency from 66 MHz to 33 MHz at t_2. The black dots represent synchronization points and the black lines simulated executions of code sections on that cores. Changing the clock frequency might also be initiated by an external event or by a simulated power management hardware unit at particular points in simulated time. Before frequencies of cores can be changed, synchronization is required. Core 1 enters a synchronization point at t_0 and waits until core 0 and core 2 have been simulated from their initial till

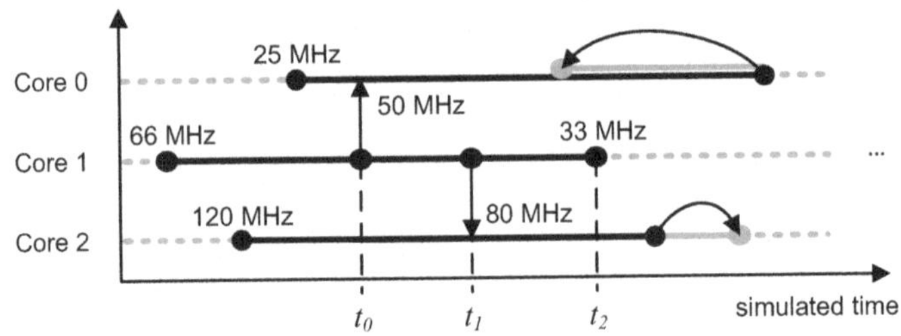

Fig. 3.12 Illustration of our dynamic frequency adaption method. The activity running on core 1 dynamically changes the frequencies of core 0, core 2, and its own core by calling a function CHANGECOREFREQUENCY, see Algorithm 3.2

Algorithm 3.2 Updates the clock frequency of a given core and is called after a simulation task entered a synchronization point.

function CHANGECOREFREQUENCY($core, t_0, f_0$)
  $t_1 = getTime(core)$
  $f_1 = getFrequency(core)$
  if $t_1 - t_0 > 0$ **then**
    $\Delta t = (t_1 - t_0)\left(f_1 / f_0 - 1\right)$
    $addTime(core, \Delta t)$
    $updateMinHeap(core)$
  $setFrequency(core, f_0)$

their next synchronization points. The execution times are determined based on their initial frequencies and bandwidth contention is safely considered until t_0. Now, the simulation of core 1 is continued and the frequency of core 0 is changed in its past from 25 to 50 MHz. Thus, the execution time of core 0 has to be re-calculated for the interval of t_0 and its second synchronization point assuming a clock frequency of 50 instead of 25 MHz. The clock speed gets doubled and therefore, the execution time gets halved for this interval. The synchronization point of core 0 is moved into the past and its position in the global min-heap is updated. Since core 1 still has the smallest timestamp, its simulation is continued until t_1 and the frequency of core 2 gets decreased from 120 to 80 MHz. The execution time of core 2 is increased accordingly, its synchronization point is moved into the future, and its position in the min-heap is updated. At t_2, core 1 changes its own frequency after synchronization happened. No re-calculation of the execution time is required, because the new frequency is only valid from now for the next and not for the last code sections. A detailed description of our frequency adaption method is given in Algorithm 3.2. This method allows to develop and evaluate new DPM techniques for reducing power consumption or for shaping activity execution times in InvadeSIM.

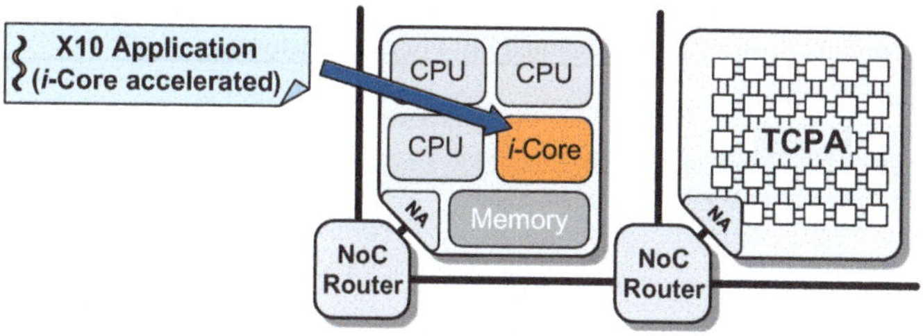

Fig. 3.13 Simulation of an activity running on and being accelerated by an *i*-Core

3.3.5 *Modeling the* i-*Core*

In the last sections, the simulation and timing estimation for general-purpose processors has been presented. However, heterogeneous architectures such as used in invasive computing may also contain special-purpose processors. Therefore, in this section, we introduce a timing simulation model for the run-time adaptable *i*-Core [22] (see Fig. 3.13). This model does not include a detailed cycle-accurate simulation of the internal reconfigurable structure of the *i*-Core, but only a high-level simulation of the timing effects an activated *i*-Core acceleration provides. Furthermore, an API at X10-level is provided to load and control the execution of different accelerators. This low-level interface is used by the invasive runtime system to realize the invasion and infection of *i*-Cores.

The basic idea of this special-purpose core is that it allows extending its ISA by special instructions on demand through dynamic reconfiguration. Such special instructions are implemented by individual hardware accelerators, which are loaded into the reconfigurable fabric of the *i*-Core and are internally connected to its microarchitecture. If a special instruction is then used by the application, it is executed much faster than if it would have used a sequence of regular instructions instead. The speedup is typically determined in average for a specific code section using special instructions, not only for a single instruction itself. If an application does not use *i*-Core accelerated instructions, the *i*-Core behaves like a normal RISC core and any application code can be executed on it.

Listing 3.1 InvadeSIM *i*-Core application programming interface.

```
class ICoreDriver {
    static def isOnICore() : boolean;
    static def loadAccelerator(id:int) : void;
    static def startAcceleratedPhase(id:int, speedup:double) :
          void;
    static def stopAcceleratedPhase() : void;
}
```

Listing 3.2 Example of an *i*-Core accelerated fast Fourier transform.

```
class ICoreExample {
    def runFFT_ICore(signal:Rail[double], spectrum:Rail[double]) {
        if (ICoreDriver.isOnICore()) {
            // Load FFT accelerator
            ICoreDriver.loadAccelerator(ICoreDriver.FFT_ID);
            ICoreDriver.startAcceleratedPhase(ICoreDriver.FFT_ID, 5.2);
            runFFT(signal, spectrum);
            ICoreDriver.stopAcceleratedPhase();
        } else {
            runFFT(signal, spectrum);
        }
    }
}
```

To simulate *i*-Core specific behavior, we provide an X10 API, which allows selecting different accelerators as well as to mark code sections for accelerated execution. In Listing 3.1, the main functions of the API are presented. The first method `isOnICore` returns, whether or not the current activity is executed on an *i*-Core. Depending on the result, different code sections may be executed, one using special instructions for the *i*-Core and one using only regular instructions as a fallback solution. The method `loadAccelerator` virtually loads the accelerator with the identifier `id` into the reconfigurable fabric of the *i*-Core. In our simulation model, the accelerator id is stored in the respective *i*-Core hardware data object and the required simulated time for reconfiguration can be configured. This method is typically called during an invasion phase, if the resource is available. Application activities can be accelerated only after an accelerator has been virtually loaded. The application programmer has to mark an *i*-Core code section using `startAcceleratedPhase` and `stopAcceleratedPhase` and has to provide the accelerator id and the speedup measured for this code section. This information is stored in the simulation task of the activity. In fact, the timing of this code section is determined by our direct-execution simulation approach and no special instructions are simulated for it. Instead, the execution time $t_{i,j}^{icore}$ is calculated by dividing $t_{i,j}$ from Eq. (3.1) by the given speedup, if the provided accelerator id matches with the id from the loaded accelerator:

$$t_{i,j}^{icore} = t_{i,j} / speedup \tag{3.4}$$

If the activity is re-scheduled by the invasive runtime system after being blocked to a RISC core or to another *i*-Core without the required accelerator being loaded, regular execution time calculation is applied. If a deterministic execution of the activity always on an appropriate *i*-Core is desired, the InvadeSIM OctoPOS API allows pinning an activity to a specific core.

In Listing 3.2, a code example of the *i*-Core API usage is depicted. Assuming an accelerator for a *fast Fourier transform* (FFT) calculation is available for the *i*-Core, this code example shows how to wrap an existing FFT calculation to be accelerated by an *i*-Core if available or to provide a fallback solution if not. Compared to a regular execution of this FFT transforming an input signal from time domain (`signal`) into frequency domain (`spectrum`) on a RISC core, an exemplary speedup of 5.2 for an execution on an *i*-Core is assumed. If the activity is running on an *i*-Core (line 3), the accelerator for the FFT is loaded (line 5) and the code section is marked with the specified speedup of 5.2 (line 6). Now, `runFFT` is called and the timing and functionality is simulated by our direct-execution simulation approach. In contrast to a normal execution on a RISC core, the execution time on the *i*-Core is determined by Eq. (3.4). Once the FFT is calculated, the speedup marker is removed (line 8). If the activity is not executed on an *i*-Core, `runFFT` is called without acceleration (line 10).

3.3.6 Modeling the TCPA

Besides acceleration through run-time adaptable microarchitectures, invasive computing also investigates a special class of massively parallel accelerators called *tightly-coupled processor arrays* (TCPAs) (see Fig. 3.14). These are highly optimized programmable processor arrays, specifically designed for an energy-efficient and timing-predictable execution of computationally intensive applications that can be expressed by nested loops. Areas of application are among others scientific computing, robotics, image, and signal processing. In this section, we introduce a timing simulation model for TCPAs. Similar to the *i*-Core timing model, no simulation of architectural details is provided, but a high-level timing estimation based on latency expressions. Furthermore, an X10 API is provided to virtually load and execute programs on TCPAs in InvadeSIM. The invasive runtime system uses this low-level interface to realize invasion and infection of TCPAs.

A TCPA consists of an $N \times M$ array of weakly programmable PEs each having a configurable VLIW structure. In Fig. 3.14, a tile containing a 5×6 array is depicted. The PEs are tightly coupled to each other by point-to-point connections, which allows data transfers within a couple of cycles. Not shown in the figure, the array is surrounded by buffers to feed the PEs with data and to collect the results after

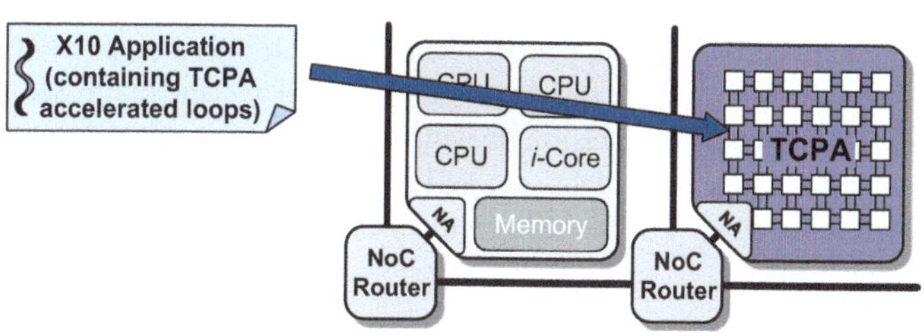

Fig. 3.14 Simulation of a compute-intensive X10 program containing nested loops being accelerated by a TCPAs

Listing 3.3 Example of a FIR filter implementation in X10.

```
class FirFilter {
    def run(a:Array[int], u:Array[int], y:Array[int], N1:int, N2:int) {
        for(var i:int = 0; i < N1; i++) {
            y(i) = 0;
            for(var j:int = 0; j < N2; j++) {
                if (i - j >= 0) {
                    y(i) += a(j) * u(i-j);
                } else {
                    u(i) = 0;
                }
            }
        }
    }
}
```

the computations are finished. Also not shown, a RISC core that is responsible for orchestrating an execution on the TCPA by filling the buffers with data, starting the array, waiting for the computations are finished, and finally reading the data back from the buffers into a larger memory such as the TLM. Large datasets are split up into chunks and processed separately, while TCPA computations and memory transfers are overlapped in order to provide a continuous stream of data to the array and prevent the array from stalling.

In order to map applications onto a TCPA, the authors of [23, 24] have developed a methodology to symbolically parallelize and schedule nested loop programs onto such massively parallel architectures, which is briefly summarized in this paragraph. Such programs consist of several nested loops, which iterate over the elements of the input data and apply calculations in their loop body to produce the corresponding output data. An example of such a program is an FIR filter, which can be described by the simple difference equation $y(i) = \sum_{j=0}^{N-1} a(j) \cdot u(i-j)$ with $0 \leq i \leq T-1$, where N denotes the number of filter taps, T denotes the number of samples over time, $a(j)$ are the filter coefficients, $u(i)$ are the filter inputs, and $y(i)$ are the filter results. In Listing 3.3, an X10 implementation of such an FIR filter is shown. The 2-dimensional iteration space of the nested loop has a size of $N_1 \cdot N_2$ and is partitioned by the symbolic mapping approach into 2-dimensional tiles having an equal size[5] of $p_1 \cdot p_2$ according to the number of available PEs in both dimensions, P_1 and P_2. At run time, each tile is then mapped to one PE, which only computes its fraction of the full iteration space. Now, the parameters P_1 and P_2 (the claim size) become only known at run time after invasion as well as the iteration space bounds N_1 and N_2, which depend on the input data size and might change for each infection (e.g., images of different sizes). The resulting tile sizes p_1 and p_2 are determined by the following equations:

[5] Perfect tiling is assumed.

Table 3.2 Symbolic latency expressions for different applications targeting TCPAs. Derived from the results presented in [23]

Application	Latency expressions
FIR filter	$L_1 = p_1(P_1 - 1) + (1 - p_1 + p_1 p_2)(P_2 - 1) + p_1 p_2$
	$L_2 = (1 - p_2 + p_1 p_2)(P_1 - 1) + p_2(P_2 - 1) + p_1 p_2$
Matrix multiplication	$L_1 = p_1(P_1 - 1) + (1 - p_1 + p_1 p_2)(P_2 - 1) + (1 - p_1 p_2 + p_1 p_2 p_3)(P_3 - 1) + p_1 p_2 p_3$
	$L_2 = (1 - p_2 p_3 + p_1 p_2 p_3)(P_1 - 1) + p_2(P_2 - 1) + (1 - p_2 + p_2 p_3)(P_3 - 1) + p_1 p_2 p_3$
	$L_3 = (1 - p_2 p_3 + p_1 p_2 p_3)(P_1 - 1) + (1 - p_3 + p_2 p_3)(P_2 - 1) + p_3(P_3 - 1) + p_1 p_2 p_3$
	$L_4 = (1 - p_2 + p_1 p_2)(P_1 - 1) + p_2(P_2 - 1) + (1 - p_1 p_2 + p_1 p_2 p_3)(P_3 - 1) + p_1 p_2 p_3$
	$L_5 = (1 - p_3 + p_1 p_3)(P_1 - 1) + (1 - p_1 p_3 + p_1 p_2 p_3)(P_2 - 1) + p_3(P_3 - 1) + p_1 p_2 p_3$
	$L_6 = p_1(P_1 - 1) + (1 - p_1 p_3 + p_1 p_2 p_3)(P_2 - 1) + (1 - p_1 + p_1 p_3)(P_3 - 1) + p_1 p_2 p_3$

$$p_1 = \left\lceil \frac{N_1}{P_1} \right\rceil \quad p_2 = \left\lceil \frac{N_2}{P_2} \right\rceil \tag{3.5}$$

The symbolic mapping approach allows mapping applications on an array of PEs dynamically, where the number of used PEs is only known at run time. Due to a timing-predictable operation principle of the TCPA, it also allows predicting the latency L exactly in terms of required clock cycles of a program in dependence on the number of claimed PEs and the input data size. In [23], the authors describe how to derive latency expressions from schedule vectors, which indicate at which clock cycle an iteration of a tile is computed. In Table 3.2, the latency expressions for an FIR filter and a matrix multiplication are depicted. Depending on the dimensionality n of an application, multiple schedule vectors respectively latency expressions exist due to different scanning orders along the dimensions. According to [23], the number of scanning orders and consequently latency expressions for an n-dimensional loop program is given by $n!$. Before an application is started at run time, a prologue selects the optimal schedule vector resulting in the smallest latency in dependence on the concrete values for the symbolic parameters.

Instead of a complex cycle-accurate simulation of the TCPA, we developed a high-level timing model of the TCPA using the introduced latency expressions rather than simulating instruction-by-instruction of arrays easily reaching several 100s of PEs, and emulate the functionality of the programs by executing regular X10 code. In order to control the TCPA simulation model in InvadeSIM, we provide an X10 API, which is depicted in Listing 3.4. The method `hasTcpa` checks, whether the current tile contains a TCPA. Similar to the loading of an accelerator into the i-Core, the TCPA requires loading configuration code into the PEs. The method `loadConfig` virtually loads a configuration identified by `id` to the specified PEs defined by the

Listing 3.4 InvadeSIM TCPA application programming interface.

```
class TcpaDriver {
    static def hasTcpa() : boolean;
    static def loadConfig(id:int, P:Rail[int]) : void;
    static def fillBuffers(size:int) : void;
    static def startTcpa(N:Rail[int], behavior:()=>void) : void;
    static def emptyBuffers(size:int) : void;
}
```

one-dimensional array P. The parameter P contains the number of PEs in each dimension, which are going to be configured. P(0) and P(1) would correspond to the already introduced parameters P_1 and P_2 in case of the FIR filter. These numbers are limited by the simulated physical bounds of N rows and M columns of the TCPA. For example, a value of [2,3] for this parameter would configure a 2×3 array of PEs. A configuration is specific to an application and contains the respective latency expressions. They are initialized during system initialization as exemplary depicted in Listing 3.5 for the FIR filter. As shown in line 2 and 4, the two latency expressions from Table 3.2 are represented by X10 *anonymous functions*, also called *lambda expressions*. They require the two parameters p and P, which correspond to the tile sizes and number of PEs for each dimension and return the respective latency. They are added to the FIR filter configuration and later used to determine the smallest latency. A virtual size of the configuration in bytes can be provided, which also allows simulating the timing of fetching a configuration from a global memory tile. The methods fillBuffers and emptyBuffers simulate the timing of transferring an amount of size bytes to or from the TCPA buffers. During the simulated time of communication, the calling activity is blocked. The method startTcpa virtually starts the configured TCPA computation and waits until it is finished. Direct-execution simulation is stopped and the required execution time t^{tcpa} for a specific application being accelerated by the TCPA is determined by:

$$t^{tcpa} = \min_{\forall i} \{L_i\} \frac{1}{f^{tcpa}} \tag{3.6}$$

Listing 3.5 Creating an FIR filter configuration as part of the TCPA API initialization.

```
1    val config = new TcpaConfiguration("FIR_FILTER", 32768);
2    val e1 = (p:Rail[long], P:Rail[int]) =>
3            p(0)*(P(0)-1) + (1-p(0)+p(0)*p(1))*(P(1)-1) + p(0)*p(1);
4    val e2 = (p:Rail[long], P:Rail[int]) =>
5            (1-p(1)+p(0)*p(1))*(P(0)-1) + p(1)*(P(1)-1) + p(0)*p(1);
6    config.addLatencyExpression(e1);
7    config.addLatencyExpression(e2);
8    TcpaDriver.initConfig(TcpaDriver.FIR_FILTER_ID, config);
```

Listing 3.6 Example of a TCPA accelerated FIR filter computation.

```
1 class TcpaExample {
2     val N1 = 1024;
3     val N2 = 768;
4     def runFirFilter_Tcpa(a:Array[int], u:Array[int], y:Array[int]) {
5         if (TcpaDriver.hasTcpa()) {
6             // Load FIR filter configuration for a 2x3 array
7             TcpaDriver.loadConfig(TcpaDriver.FIR_FILTER_ID, [2, 3]);
8             TcpaDriver.fillBuffers((a.size + b.size) * 4);
9             TcpaDriver.startTcpa([N1,N2], ()=>{FirFilter.run(a,u,y,N1,N2)
                  ;});
10            TcpaDriver.emptyBuffers(c.size * 4);
11        } else {
12            FirFilter.run(a,u,y,N1,N2);
13        }
14    }
15 }
```

The application provides the set of latency expressions L_i as exemplary depicted in Eq. 3.2, which are then evaluated using the parameters `p` and `P` determined at run time. The tile sizes `p` are calculated according to Eq. 3.5 given the iteration space bounds `N` and the number of PEs in each dimension `P`. To simulate the functionality of the TCPA computations, the anonymous function `behavior` is called. Any X10 code can be executed and it does not influence the TCPA execution time. Afterwards, the simulated time is advanced by t^{tcpa}, the calling activity is unblocked, and the direct-execution simulation is continued.

In Listing 3.6, a code example of the TCPA API usage is provided. An FIR filter is supposed to be accelerated. An appropriate configuration is existing and indicated by `FIR_FILTER_ID`. Assuming, a 2×3 array of PEs on a TCPA tile has already been invaded. If the calling activity is executed on this tile, the FIR filter configuration is loaded to all available PEs in line 8. Filling the buffers with the input signal `a` and the filter coefficients `b` is simulated in line 9. Afterwards, the simulation of the TCPA timing and functionality is applied in line 10 given the iteration space bounds `N1` and `N2` of the input signal and the behavioral function of the FIR filter. Once this is finished, transferring the result array `c` from the buffers into the TLM is simulated in line 11. If this activity is executed on a tile with no TCPA, the FIR filter is executed regularly on a RISC core and simulated by our direct-execution simulation approach. In both cases, the same results are produced, but the execution times differ.

3.4 Related Work

Many different techniques for simulating programmable devices have been evolved over decades. In this section, a summary of these techniques including popular representatives as well as some definitions of important terms in this area are discussed.

The first simulators for microprocessors are as old as programmable devices themselves and have been developed by hardware designers. These are *micro-architecture simulators* or *cycle-accurate simulators*. Micro-architecture simulators are either written in a hardware description language such as VHDL or in a high-level programming language such as C/C++. They model the microarchitecture of a processor including the instruction pipeline with its components such as ALU, cache memory, or branch prediction at register-transfer level (RTL). Since such simulators provide cycle-accurate timing information, they allow designers to evaluate the performance and efficiency of novel architectural features without a costly and time-consuming fabrication of a prototyping chip. However, they have a slowdown of around 20,000, which allows them simulating a very short sequence of instructions only and makes it almost impossible to use them for full-system simulation. In academia, well known single-processor simulators are SimpleScalar [25], PTLsim [26], and SWARM [27]. Some of them serve as the basis for multi-processor simulators. For instance, MPARM [28], an MPSoC simulator for ARM processors, was derived from SWARM. Other cycle-accurate simulators for multi-processors are RSIM [29] and gem5 [30], a merge of M5 [31] for full-system simulation and GEMS [32] for a cache-coherent memory system. Due to their detailed simulation models and long simulation times, higher abstraction levels as well as faster simulation techniques have been developed, which will be discussed in the following.

Other types of processor simulators are *macro-architecture simulators* and *instruction set simulators (ISS)*. They model the behavior of the instructions of a target processor without a detailed pipeline model and maintain the target machine state such as the content of registers and cache behavior. Popular languages for realizing ISS are C/C++, SystemC, and Java. They allow simulating a target assembly program instruction-by-instruction. The target program is stored in the host memory and the simulator operates like a processor in a *fetch-decode-execute loop*. It fetches an instruction from memory, decodes it and depending on the opcode calls a routine that simulates the effect of that opcode. They are a very important step in the development of efficient computer architectures, because they allow testing real programs on the target architecture in interaction with other components of the hardware design. Because they model fewer details, they have slowdowns of around 1,000 and can execute much longer instruction streams than cycle-accurate simulators. Therefore, they are used for studying cache performance and debugging operating system code or compilers. Such ISS are also called *interpretive ISS* [25, 33–35], because they interpret each instruction at run time. This technique allows the highest degree of flexibility, however, decoding each instruction at run time is a very time-consuming process in software simulation.

Thus, a couple of optimizations have been proposed to improve the performance of interpretive ISS. One of them is *compiled ISS* [36–39]. Here, the time-consuming instruction decoding phase is moved from run time to compile time. This technique improves the simulation performance significantly, however, at the price of flexibility. All compiled ISS assume that the target program instructions are known at compile time, which excludes them from being used in many application domains. Especially for systems with run-time dynamic program code, as provided by operating systems, compiled ISS cannot be used. Thus, *cache-compiled ISS* [40–43] were introduced, which are almost as fast as compiled ISS, but keep the flexibility of interpretive ISS. The basic simulation loop is the same as in interpretive ISS. However, instead of decoding each instruction at each occurrence, an instruction is only decoded at its first occurrence or after a modification. Its address as well as a pointer to the function that simulates the behavior of that instruction and manipulates the target machine state is put into a cache data structure, which is consulted at subsequent instruction occurrences. Commercial examples of ISS are Intel's Simics [44], ARM's ARMulator [45], Synopsis' nSIM [46], and Imperas' ISS [47]. Academic representatives are SPIM [48] and Sulima [49].

Other research directions address the efficient retargetability of ISS using architecture description languages (ADLs) where a simulator, compiler, assembler, and debugger are automatically generated based on a specification of the processor architecture [34, 37, 50–53].

In the literature, also hybrid simulation techniques have been investigated to accelerate ISS. Here, the simulator allows switching between native code execution (fast-forwarding mode) for higher simulation rates at the cost of reduced accuracy and ISS-based execution for a detailed yet slow simulation at run time. This technique allows a hardware designer or software developer to trade-off simulation performance for accuracy depending on his requirements. For example, debugging a complex application does only require to simulate that part in detail where the problem is expected, the other parts such as booting up the operating system may be simulated less detailed yet much faster. A popular representative is HySim [54–56].

Another technique to speed up instruction set simulation is *dynamic binary translation* [57, 58]. This techniques is similar to cache-compiled ISS. The difference is the granularity of a translation unit. In case of cache-compiled ISS, one instruction is decoded and translated into host instructions, whereas in dynamic binary translation a complete basic block is translated into a sequence of host instructions. Equal to both techniques is that they operate on a simulated state (e.g., register file and memory content) rather than on the real machine state and that they use caching techniques to avoid retranslations of the same instructions or blocks. Popular classic simulators based on this technique are FastSim [59], Embra [60], and Shade [61]. Other simulators based on dynamic binary translation are Imperas' OVPsim [62], AMD's SimNow [63], SimOS [64], and the emulator QEMU [65].

In order to cope with the increased complexity of embedded MPSoCs, designing such systems on system level became crucial. Tools have been developed to create entire software-based virtual platforms including different models of processors, hardware peripherals, accelerators as well as communication infrastructures. In this area of platform design for SoCs, the following commercial framework are well known: Simics [44] from Intel, CoMET, METeor, and Platform Architect from Synopsys [66], as well as CoFluent Studio [67] from CoFluent Design, Magillem Platform Assembly [68] from Magillem Design Services and the Open Virtual Platform (OVP) from Imperas [69]. They are software development environments, which allow for flexible design and high-speed simulation of processor-based systems. However, these tools are mainly targeted at virtual prototyping to support early software development before real hardware is available. The effort to build and update the underlying models is quite high and thus, they are less suited for exploratory investigations where in most cases a full application model is not yet available and fast modifications of both application and hardware architecture are a key issue. Also, the integration of custom components, e.g., dedicated accelerators such as TCPAs, into the aforementioned frameworks could be a difficult task since accuracy and performance have to be weighed up and a mixture of functional abstractions have to be supported when simulating resource-aware applications.

In embedded system design, there exist several source level simulation methods for software simulation at system level [70–76]. These approaches rely on the source code instrumentation, i.e., the source code of the applications is back-annotated with timing information obtained by low-level simulators. Such simulation approaches also make use of TLM-based virtual prototypes, containing multiple processors [1, 2]. The aforementioned concepts are also often called *host-compiled simulation* and are closely related to our presented method since they use a source code functional representation. However, the major difference to our approach is that we do not need to generate an intermediate source code and have to perform a timing analysis for each new application scenario before the actual simulation, but we use a simulation approach that is directly based on time measurement and a subsequent on-the-fly time-warping mechanism. So, our approach can be seen as a compromise between accuracy and performance in order to evaluate different concepts of resource-aware programming as well as resource management on various architectures that are not fully available yet.

In the domain of HPC, software simulation techniques for performance evaluation have a long tradition as in the domain of embedded systems. The target architectures studied in HPC are highly parallel processing systems, which have similarities to our considered target architectures. It is of high interest of architects to study the interactions of parallel programs and the parallel architectures executing them to compare different organizations without building expensive hardware prototypes. However, simulating large problems comprising parallel machines with realistic workloads requires a huge amount of computation and memory. Thus, besides different workload models (e.g., distribution-driven, trace-driven, or execution-driven workload characterizations), mainly two techniques have been studied to accelerate simulation: *direct execution* and *parallel simulation*. Parallel simulation speeds up simulation by exploiting parallelism inherently given by the parallel target machine and the parallel host computer. Due to significant contributions to parallel simulation in this book, this topic will be discussed separately in Chap. 5. In direct execution, a parallel program from the target system is directly executed on the simulation host machine. Timing information about the program's execution is generated dynamically based on its execution on the host machine. Classic direct-execution simulators for single and multi-processors are the Rice Parallel Processing Testbed [11], the Wisconsin Wind Tunnel I [12] and II [13]. The presented simulation techniques are very related to those proposed in this book. However, the major difference compared to our simulation approach is the determination of the target simulation time based on a tight interaction of performance counter predictors and an intra-tile shared memory contention model.

3.5 Summary

In this chapter, we presented novel concepts of direct-execution simulation, which are used for the performance evaluation of resource-aware parallel applications on heterogeneous MPSoC architectures. Using this simulation approach, we are able to simulate the functional behavior of parallel applications as well as important timing information about their execution on simulated heterogeneous target architectures. Furthermore, we can easily change the parameters of the simulated architecture in terms of computational power of the containing processing resources and memory bandwidths. Altogether, it allows us to evaluate the performance and the functional correctness of the applications on a heterogeneous architecture in almost native execution time. The simulation approach is composed of two parts. These are (a) a performance counter measurement on the host machine and a time-warping mechanism for scaling the counted events to a modeled target processor and (b) a synchronization mechanism for simulating different simulation tasks. Timing models for the simulation of invasive hardware accelerators such as TCPAs and *i*-Cores have been presented as well as a shared-memory bandwidth contention model and a dynamic frequency adaption technique. Experiments using an X10 benchmark showed a maximum slowdown of 5 for single-core simulation compared to native execution, while resulting in an error of 18%.

References

1. S. Stattelmann, O. Bringmann, W. Rosenstiel, Fast and accurate resource conflict simulation for performance analysis of multi-core systems, in *Proceedings of the Conference on Design, Automation and Test in Europe (DATE)*, Grenoble, France, 14–18 Mar 2011 (IEEE, 2011), pp. 210–215. ISBN: 978-1-61284-208-0. https://doi.org/10.1109/DATE.2011.5763044
2. A. Gerstlauer, Host-compiled simulation of multi-core platforms, in *Proceedings of the 21st IEEE International Symposium on Rapid System Prototyping (RSP)*, Fairfax, VA, USA, 8–11 June 2010 (IEEE, 2010), pp. 1–6. ISBN: 978-1-4244-7073-0. https://doi.org/10.1109/RSP.2010.5656352
3. S. Ha J. Teich (eds.), *The Handbook of Hardware/Software Codesign* (Springer, Netherlands, 2017). ISBN: 978-94-017-7266-2. https://doi.org/10.1007/978-94-017-7267-9
4. T. Wild, A. Herkersdorf, G.-Y. Lee, TAPES-tracebased architecture performance evaluation with system C. Design Autom. Embedded Syst. **10**(2–3), 157–179, (Sept. 2005). ISBN: 0929-5585. https://doi.org/10.1007/s10617-006-9589-4
5. R.E. Wunderlich et al., Statistical sampling of microarchitecture simulation. ACM Trans. Model. Comput. Simul. (TOMACS) **16**(3), pp. 197–224, (July 2006). ISSN: 1049-3301. https://doi.org/10.1145/1147224.1147225
6. A. Falcon, P. Faraboschi, D. Ortega, Combining simulation and virtualization through dynamic sampling, in *Proceedings of the IEEE International Symposium on Performance Analysis of Systems & Software (ISPASS)*, San Jose, CA, USA, 25–27 Apr 2007 (IEEE, 2007), pp. 72–83. ISBN: 1-4244-1081-9. https://doi.org/10.1109/ISPASS.2007.363738
7. F. Hannig et al., Resource-aware programming and simulation of MPSoC architectures through extension of X10, in *Proceedings of the 14th International Workshop on Software and Compilers for Embedded Systems (SCOPES)*, St. Goar, Germany, 27–28 June 2011, (ACM, 2011), pp. 48–55. ISBN: 978-1-4503-0763-5. https://doi.org/10.1145/1988932.1988941

8. S. Roloff, F. Hannig, J. Teich, Approximate time functional simulation of resource-aware programming concepts for heterogeneous MPSoCs, in *Proceedings of the 17th Asia and South Pacific Design Automation Conference (ASP-DAC)*, Sydney, NSW, Australia, 30 Jan–2 Feb, 2012 (IEEE, 2012), pp. 187–192. ISBN: 978-1-4673-0770-3. https://doi.org/10.1109/ASPDAC.2012.6164943
9. S. Roloff, F. Hannig, J. Teich, Fast architecture evaluation of heterogeneous MPSoCs by host-compiled simulation, in *Proceedings of the 15th International Workshop on Software and Compilers for Embedded Systems (SCOPES)*, St. Goar, Germany, 15–16 May 2012 (ACM, 2012), pp. 52–61. ISBN: 978-1-4503-1336-0. https://doi.org/10.1145/2236576.2236582
10. C.J. Mauer, M.D. Hill, D.A. Wood, Full-system timing first simulation, in *Proceedings of the ACM SIGMETRICS International Conference on Measurement and Modeling of Computer Systems (SIGMET RICS)*, Marina Del Rey, CA, USA, 15–19 June 2002 (ACM, 2002), pp. 108-116. ISBN: 1-58113-531-9. https://doi.org/10.1145/511334.511349
11. R.C. Covington et al., The rice parallel processing Testbed, in *Proceedings of the ACM SIGMETRICS Conference on Measurement and Modeling of Computer Systems (SIGMETRICS)*, Santa Fe, NM, USA, 24–27 May 1988 (ACM, 1988), pp. 4–11. ISBN: 0-89791-254-3. https://doi.org/10.1145/55595.55596
12. S.K. Reinhardt et al., The wisconsin wind tunnel: virtual prototyping of parallel computers, in *Proceedings of the ACM SIGMETRICS Conference on Measurement and Modeling of Computer Systems (SIGMETRICS)*, Santa Clara, CA, USA, 10–14 May 1993 (ACM, 1993), pp. 48–60. ISBN: 0-89791-580-1. https://doi.org/10.1145/166955.166979
13. S.S. Mukherjee et al., Wisconsin wind tunnel II: A fast, portable parallel architecture simulator, in *IEEE Concurrency 8.4 (Oct. 2000)*, pp. 12–20. ISSN: 1092-3063. https://doi.org/10.1109/4434.895100
14. G. Zheng, G. Kakulapati, L. Kalè, BigSim: a parallel simulator for performance prediction of extremely large parallel machines, in *Proceedings of the 18th International Parallel and Distributed Processing Symposium (IPDPS)*, Santa Fe, NM, USA, 26–30 Apr 2004 (IEEE, 2014), pp. 78–87. ISBN: 0-7695-2132-0. https://doi.org/10.1109/IPDPS.2004.1303013
15. T. Diop, N.E. Jerger, J. Anderson, Power modeling for heterogeneous processors, in *Proceedings of the 7th Workshop on General Purpose Processing Using GPUs (GPGPU)*, Salt Lake City, UT, USA, 1 Mar 2014 (ACM, 2014), 90:90-90:98. ISBN: 978-1-4503-2766-4. https://doi.org/10.1145/2576779.2576790
16. R. Bertran et al, Counter-based power modeling methods: top-down vs. bottom-up. Comput. J. **56**(2), pp. 198–213 (Feb. 1, 2013). ISSN: 1460-2067. https://doi.org/10.1093/comjnl/bxs116
17. K. Singh, M. Bhadauria, S.A. McKee, Real time power estimation and thread scheduling via performance counters. ACM SIGARCH Comput. Archit. News **37**(2), 46–55 (May 2009). ISSN: 0163-5964. https://doi.org/10.1145/1577129.1577137
18. G. Contreras, M. Martonosi, Power prediction for intel XScale ® processors using performance monitoring unit events, in *Proceedings of the International Symposium on Low Power Electronics and Design (ISPLED)*, San Diego, CA, USA, 8–10 Aug 2005 (ACM, 2005), pp. 221–226. ISBN: 1-59593-137-6. https://doi.org/10.1145/1077603.1077657
19. J. Russ, M. Margaret, Run-time power estimation in high performance microprocessors, in *Proceedings of the International Symposium on Low Power Electronics and Design (ISPLED)*, Huntington Beach, CA, USA, 6–7 Aug 2001 (ACM, 2001), pp. 135–140. ISBN: 1-58113-371-5. https://doi.org/10.1109/LPE.2001.945389
20. S. Eranian, Perfmon2: a flexible performance monitoring interface for Linux, in *Proceedings of the Ottawa Linux Symposium (OLS)*, Ottawa, ON, Canada. 19–22 July 2006, pp. 269–288
21. V.M. Weaver, Linux perfevent features and overhead, in *Proceedings of the 2nd International Workshop on Performance Analysis of Workload Optimized Systems (FastPath)*. Austin, TX, USA. 21 Apr 2013, pp. 80–86
22. J. Henkel et al., i-Core: a run-time adaptive processor for embedded multi-core systems, in *Proceedings of the International Conference on Engineering of Reconfigurable Systems and Algorithms (ERSA)*, Las Vegas, NV, USA, July 18–21, 2011, pp. 1–8

23. J. Teich, A. Tanase, F. Hannig, Symbolic mapping of loop programs onto processor arrays. J. Signal Process. Syst. (JSPS) **77**(1–2), pp. 31–59 (Oct. 2014). ISSN: 1939-8018. https://doi.org/10.1007/s11265-014-0905-0
24. A. Tanase et al., Symbolic multi-level loop mapping of loop programs for massively parallel processor arrays. ACM Trans. Embedded Comput. Syst. (TECS) **17**(2), 31:1–31:27 (Apr. 2018). ISSN: 1539-9087. https://doi.org/10.1145/3092952
25. T. Austin, E. Larson, D. Ernst, SimpleScalar: an infrastructure for computer system modeling. Computer **35**(2), 59–67 (Aug. 7, 2002). ISSN: 0018-9162. https://doi.org/10.1109/2.982917
26. M.T. Yourst, PTLsim: a cycle accurate full system x86-64 microarchitectural simulator, in *Proceedings of the IEEE International Symposium on Performance Analysis of Systems & Software (ISPASS)*, San Jose, CA, USA, 25–27 Apr 2007 (IEEE, 2007), pp. 23–34. ISBN: 1-4244-1081-9. https://doi.org/10.1109/ISPASS.2007.363733
27. M. Dales, SWARM 0.44 Documentation. Tech. rep. Department of Computer Science, University of Glasgow, Nov. 10, 2000. http://www.cl.cam.ac.uk/~mwd24/phd/swarm.html
28. L. Benini et al., MPARM: exploring the multi-processor SoC design space with system C. J. VLSI Signal Process. Syst. Signal Image Video Technol. **41**(2), 169–182 (Sept. 1, 2005). ISSN: 0922-5773. https://doi.org/10.1007/s11265-005-6648-1
29. C.J. Hughes et al., RSIM: simulating shared-memory multiprocessors with ILP processors. Computer **35**(2), 40–49 (Aug. 7, 2002). ISSN: 0018-9162. https://doi.org/10.1109/2.982915
30. N. Binkert et al., The Gem5 simulator. ACM SIGARCH Comput. Architect. News **39**(2), 1–7 (May 2011). ISSN: 0163-5964. https://doi.org/10.1145/2024716.2024718
31. N.L. Blinkert et al., The M5 simulator: modeling networked systems. IEEE Micro **26**(4), 52–60 (Sept. 21, 2006). ISSN: 0272-1732. https://doi.org/10.1109/MM.2006.82
32. M.M.K. Martin et al., Multifacet's general execution-driven multiprocessor simulator (GEMS) toolset. ACM SIGARCH Comput. Archit. News **33**(4), 92–99 (Nov. 2005). ISSN: 0163-5964. https://doi.org/10.1145/1105734.1105747
33. M.R. Hartoog et al., Generation of software tools from processor descriptions for hardware/-software codesign, in *Proceedings of the 34th Annual Design Automation Conference (DAC)*, Anaheim, CA, USA, 9–13 June 1997 (ACM, 1997), pp. 303–306. ISBN: 0-89791-920-3. https://doi.org/10.1145/266021.266110
34. G. Hadjiyiannis, S. Hanono, S. Devadas, ISDL: an instruction set description language for retargetability, in *Proceedings of the 34th Annual Design Automation Conference (DAC)*, Anaheim, CA, USA, 9–13 June 1997 (ACM, 1997), pp. 299–302. ISBN: 0-89791-920-3. https://doi.org/10.1145/266021.266108
35. S. Sutarwala, P.G. Paulin, Y. Kumar, Insulin: an instruction set simulation environment, in *Proceedings of the 11th International Conference on Computer Hardware Description Languages and Their Applications (CHDL)*, Ottawa, ON, 26–28 Apr 1993 (Canada. North-Holland Publishing Co., 1993), pp. 369–376. ISBN: 0-444-81641-0
36. J. Zhu, D.D. Gajski, An ultra-fast instruction set simulator, in *IEEE Transactions on Very Large Scale Integration (VLSI) Systems* 10.3 (Dec. 10, 2002), pp. 363–373. ISSN: 1063-8210. https://doi.org/10.1109/TVLSI.2002.1043339
37. J. Zhu, D.D. Gajski, A Retargetable, ultra-fast instruction set simulator, in *Proceedings of the Conference on Design, Automation and Test in Europe (DATE)*, Munich, Germany, Mar 1999 (ACM, 1999), pp. 62–69. ISBN: 1-58113-121-6. https://doi.org/10.1145/307418.307509
38. V. Zivojnovic, S. Tjiang, H. Meyr, Compiled simulation of programmable DSP architectures. J. VLSI Signal Process. Syst. Signal Image Video Technol. **16**(1), 73–80 (May 1, 1997). ISSN: 0922-5773. https://doi.org/10.1023/A:1007916402263
39. C. Mills, S.C. Ahalt, J. Fowler, Compiled instruction set simulation. Software Pract. Exp. **21**(8), 877–889 (Aug. 1, 1991). ISSN: 1097-024X. https://doi.org/10.1002/spe.4380210807
40. W.S. Mong, J. Zhu, DynamoSim: a trace-based dynamically compiled instruction set simulator, in *Proceedings of the IEEE/ACM International Conference on Computer Aided Design (ICCAD)*, San Jose, CA, USA, 7–11 Nov 2004 (IEEE, 2004), pp. 131–136. https://doi.org/10.1109/ICCAD.2004.1382557

41. M. Reshadi, P. Mishra, N. Dutt, Instruction set compiled simulation: a technique for fast and flexible instruction set simulation, in *Proceedings of the 40th Annual Design Automation Conference (DAC)*, Anaheim, CA, USA, June 2, 2003-June 6, 2002 (ACM, 2002), pp. 758–763. ISBN: 1-58113-688-9. https://doi.org/10.1145/775832.776026
42. A. Nohl et al., A universal technique for fast and flexible instruction-set architecture simulation, in *Proceedings of the 39th Design Automation Conference (DAC)*, New Orleans, LA, USA, 10–14 June 2002 (IEEE, 2002), pp. 22–27. https://doi.org/10.1109/DAC.2002.1012588
43. R.C. Bedichek, Talisman: fast and accurate multicomputer simulation, in *Proceedings of the ACM SIGMETRICS International Conference on Measurement and Modeling of Computer Systems (SIGMETRICS)*, Ottawa, ON, Canada, 15–19 May 1995 (ACM, 1995), pp. 14–24. ISBN: 0-89791-695-6. https://doi.org/10.1145/223587.223589
44. P.S. Magnusson et al., Simics: a full system simulation platform. Computer **35**(2), 50–58 (Aug. 7, 2002). ISSN: 0018-9162. https://doi.org/10.1109/2.982916
45. ARM Limited, ARM Developer Suite Debug Target Guide. Version 1.2. Sept. 1, 2017. http://infocenter.arm.com
46. Synopsys, Inc. ARC nSIM Instruction Set Simulator for ARC Processors. Version M-2017.06. Sept. 1, 2017. https://www.synopsys.com
47. Imperas Software Ltd. ISS—The Imperas Instruction Set Simulator. Sept. 1, 2017. http://www.imperas.com
48. J. Larus, SPIM: A MIPS32 Simulator. Sept. 1, 2017. http://spimsimulator.sourceforge.net
49. M. Chapman, Sulima ISA Simulator. Sept. 1, 2017. http://www.cse.unsw.edu.au
50. M. Reshadi, N. Dutt, P. Mishra, A retargetable framework for instruction-set architecture simulation. ACM Trans. Embedded Comput. Syst. (TECS) **5**(2), 431–452 (May 2006). ISSN: 1539-9087. https://doi.org/10.1145/1151074.1151083
51. E.C. Schnarr, M.D. Hill, J.R. Larus, Facile: a Language and Compiler for High-performance Processor Simulators, in *Proceedings of the ACM SIGPLAN Conference on Programming Language Design and Implementation (PLDI)*, Snowbird, UT, USA, 20–22 June 2001 (ACM, 2001), pp. 321–331. ISBN: 1-58113-414-2. https://doi.org/10.1145/378795.378864
52. R. Leupers, J. Elste, B. Landwehr, Generation of interpretive and compiled instruction set simulators, in *Proceedings of the Asia and South Pacific Design Automation Conference (ASP-DAC)*, Wanchai, Hong Kong, 18–21 Jan 1999 (IEEE, 1999), pp. 339–342. ISBN: 0-7803-5012-X. https://doi.org/10.1109/ASPDAC.1999.760028
53. A. Halambi et al., EXPRESSION: a Language for Architecture Exploration through compiler/simulator retargetability, in *Proceedings of the Conference on Design, Automation and Test in Europe (DATE)*, Munich, Germany, 9–12 Mar 1999 (IEEE, 1999), pp. 485–490. ISBN: 0-7695-0078-1. https://doi.org/10.1109/DATE.1999.761170
54. L.G. Murillo et al., Synchronization for hybrid MPSoC full-system simulation, in *Proceedings of the 49th Annual Design Automation Conference (DAC)*, San Francisco, CA, USA, 3–7 June 2012 (ACM, 2012), pp. 121–126. ISBN: 978-1-4503-1199-1. https://doi.org/10.1145/2228360.2228383
55. L. Gao et al., Multiprocessor performance estimation using hybrid simulation, in *Proceedings of the 45th Annual Design Automation Conference (DAC)*, Anaheim, CA, USA, 8–13 June 2008 (ACM, 2008), pp. 325–330. ISBN: 978-1-60558-115-6. https://doi.org/10.1145/1391469.1391552
56. S. Kraemer et al., HySim: a fast simulation framework for embedded software development, in *Proceedings of the 5th IEEE/ACM/IFIP International Conference on Hardware/Software Codesign and System Synthesis (CODES+ISSS)*, Salzburg, Austria, 30 Sept–3 Oct 2007 (IEEE, 2007), pp. 75–80. ISBN: 978-1-5959-3824-4. https://doi.org/10.1145/1289816.1289837
57. I. Böhm, B. Franke, N. Topham, Cycle-accurate performance modelling in an ultra-fast just-in-time dynamic binary translation instruction set simulator, in *Proceedings of the International Conference on Embedded Computer Systems: Architectures, Modeling and Simulation (SAMOS)*, Samos, Greece, 19–22 July 2010 (IEEE, 2010), pp. 1–10. ISBN: 978-1-4244-7936-8. https://doi.org/10.1109/ICSAMOS.2010.5642102

58. O. Almer et al., Scalable multi-core simulation using parallel dynamic binary translation, in *Proceedings of the International Conference on Embedded Computer Systems: Architectures, Modeling and Simulation (SAMOS)*, Samos, Greece, 18–21 July 2011 (IEEE, 2011), pp. 190–199. ISBN: 978-1-4577-0802-2. https://doi.org/10.1109/SAMOS.2011.6045461
59. E.C. Schnarr, J.R. Larus, Fast out-of-order processor simulation using memoization, in *Proceedings of the 8th International Conference on Architectural Support for Programming Languages and Operating Systems (ASPLOS)*, San Jose, CA, USA, 2–7 Oct 1998 (ACM, 1998), pp. 283-294. ISBN: 1-58113-107-0. https://doi.org/10.1145/291069.291063
60. E. Witchel, M. Rosenblum, Embra: fast and flexible machine simulation, in *Proceedings of the ACM SIGMETRICS International Conference on Measurement and Modeling of Computer Systems (SIGMETRICS)*, Philadelphia, PA, USA, 23–26 May 1996 (ACM, 1996), pp. 68–79. ISBN: 0-89791-793-6. https://doi.org/10.1145/233013.233025
61. B. Cmelik, D. Keppel, Shade: a fast instruction-set simulator for execution profiling, in *Proceedings of the ACM SIGMETRICS International Conference on Measurement and Modeling of Computer Systems (SIGMETRICS)*, Nashville, TN, USA, 16–20 May 1994 (ACM, 1994), pp. 128–137. ISBN: 0-89791-659-X. https://doi.org/10.1145/183018.183032
62. Imperas Software Ltd.OVP Guide to Using Processor Models. Jan. 29, 2015. http://www.imperas.com
63. B. Barnes, J. Slice, SimNow: a fast and functionally accurate AMD x86-64 system simulator, in *Tutorial at the IEEE International Symposium on Workload Characterization (IISWC)*, Austin, TX, USA. Oct. 6–8, 2005
64. M. Rosenblum et al., Complete computer system simulation: the SimOS approach.IEEE Parallel Distrib. Technol. Syst. Appl. **3**(4), 34–43 (1995). ISSN: 1063-6552. https://doi.org/10.1109/88.473612
65. F. Bellard, QEMU, a fast and portable dynamic translator, in *Proceedings of the Annual Conference on USENIX Annual Technical Conference (ATEC)*, Anaheim, CA, USA, 10–15 Apr 2005 (USENIX Association, 2005), p. 41
66. Synopsys Inc. http://www.synopsys.com
67. CoFluent Design. http://www.cofluentdesign.com
68. Magillem Design Services. http://www.magillem.com
69. Open Virtual Platforms, Imperas Software Ltd. http://www.ovpworld.org
70. Z. Wang, J. Henkel, Accurate source-level simulation of embedded software with respect to compiler optimizations, in *Proceedings of the Conference on Design, Automation and Test in Europe*, Dresden, Germany. EDA Consortium, 12–16 Mar 2012, pp. 382–387. ISBN: 978-3-9810801-8-6
71. P. Razaghi, A. Gerstlauer, Automatic timing granularity adjustment for host-compiled software simulation, in *Proceedings of the 17th Asia and South Pacific Design Automation Conference (ASP-DAC)*, Sydney, NSW, Australia, 30 Jan–2 Feb 2012 (IEEE, 2012), pp. 567–572. ISBN: 978-1-4673-0770-3. https://doi.org/10.1109/ASPDAC.2012.6165021
72. Z. Wang, K. Lu, A. Herkersdorf, An approach to improve accuracy of source-level TLMs of embedded software, in *Proceedings of the Conference on Design, Automation and Test in Europe (DATE)*, Grenoble, France, 14–18 Mar, 2011 (IEEE, 2011) pp. 216–221. ISBN: 978-1-61284-208-0. https://doi.org/10.1109/DATE.2011.5763045
73. S. Stattelmann, O. Bringmann, W. Rosenstiel, Fast and accurate source-level simulation of software timing considering complex code optimizations, in *Proceedings of the 48th ACM/EDAC/IEEE DesignAutomation Conference (DAC)*, San Diego, CA, USA, 5–10 June 2011 (ACM, 2011), pp. 486–491. ISBN: 978-1-4503-0636-2. https://doi.org/10.1145/2024724.2024838
74. Z. Wang, A. Herkersdorf, An efficient approach for system-level timing simulation of compiler-optimized embedded software, in *Proceedings of the 46th Annual Design Automation Conference (DAC)*, San Francisco, CA, USA, 26–31 July 2009 (ACM, 2009), pp. 220–225. ISBN: 978-1-6055-8497-3. https://doi.org/10.1145/1629911.1629973

75. Y. Hwang, S. Abdi, D. Gajski, Cycle-approximate retargetable performance estimation at the transaction level, in *Proceedings of the Conference on Design, Automation and Test in Europe (DATE)*, Munich, Germany, 10–14 Mar 2008 (IEEE, 2008), pp. 3–8. ISBN: 978-3-9810801-3-1. https://doi.org/10.1145/1403375.1403380
76. J. Schnerr et al., High-performance timing simulation of embedded software, in *Proceedings of the 45th Annual Design Automation Conference (DAC)*, Anaheim, CA, USA, 8–13 June 2008 (ACM, 2008), pp. 290–295. ISBN: 978-1-60558-115-6. https://doi.org/10.1145/1391469.1391543

Chapter 4
Hybrid Network-on-Chip Simulation

As discussed earlier, modern multi- and many-core architectures heavily use communication infrastructures such as NoCs in order to meet the increased bandwidth demands between the processors and for reasons of scalability. High-speed simulation techniques are required for the proper analysis of concurrency, utilization, and workload distribution of parallel applications running on such NoC-based architectures. Apart from accurate timing simulation of processor resources, it is of utmost importance also to accurately model the delays caused by the packet-based network communication in order to reliably verify performance numbers, or to identify any bottlenecks of the underlying architecture. Therefore, the first contribution of this chapter is an efficient hybrid NoC simulation approach that allows simulating communication delays equally accurate but in average much faster than on a cycle-by-cycle basis. Proposed are novel algorithmic and analytical techniques, which dynamically predict the transmission delays of messages considering the actual congestion in the NoC, routing information, packet lengths, and other parameters. According to these predictions, the simulation time may be automatically advanced in many cases, which drastically reduces the number of cycles the NoC simulator has to process. The second contribution of this chapter is the integration of the proposed simulation technique into the system-level multi-core simulator InvadeSIM. This integration allows simulating the network traffic that is inherently generated by common as well as invasive X10 programs during their parallel executions. It is explained, how communication requests are processed beginning at language level with the `at` construct over X10 runtime calls to the final NoC timing simulation. Experiments in running parallel real-world and multi-media applications on simulated NoC architectures show that we are able to achieve speedups of three orders of magnitude compared to cycle-accurate NoC simulators, while preserving a timing accuracy of above 95%. The idea and the evaluation of the hybrid NoC simulation approach have been published in [1] and the integration into the InvadeSIM in [2]. Before describing the details of the simulation technique, the underlying hardware model of the considered NoC architecture is briefly summarized.

S. Roloff et al., *Modeling and Simulation of Invasive Applications and Architectures*, Computer Architecture and Design Methodologies,
https://doi.org/10.1007/978-981-13-8387-8_4

4.1 Network-on-Chip Model

A generic invasive tiled architecture was already introduced earlier, Fig. 4.1 highlights the components of such an architecture related to the NoC. Amongst other resources such as processor cores and caches, each tile houses one NoC router (switch), which enables the tile to communicate with other tiles using packet switching. For this purpose, each router has multiple physical connections, called *links*, to its neighbors in all four cardinal directions and to its tile-local resources over the network adapter (NA). While mesh-like structures are common, note that our approach can be applied to arbitrary NoC topologies.

Each of the links is realized by two unidirectional point-to-point connections. When a message is going to be transmitted over the NoC, it is split into packets, which are further divided into fixed-length *flow control digits* (FLITs), corresponding to the amount of data that can be transported in one clock cycle over one link. Three different types of FLITs can be differentiated: each packet consists of (a) a header flit, the first flit of a packet that contains information about the route (e.g., the destination address) and defines also the routing for all following flits of the same packet, (b) a flexible number of body flits that contain the actual payload of data, and (c) the last flit, called tail flit that marks the end of the transmission flow. A typical packet is shown in Fig. 4.2.

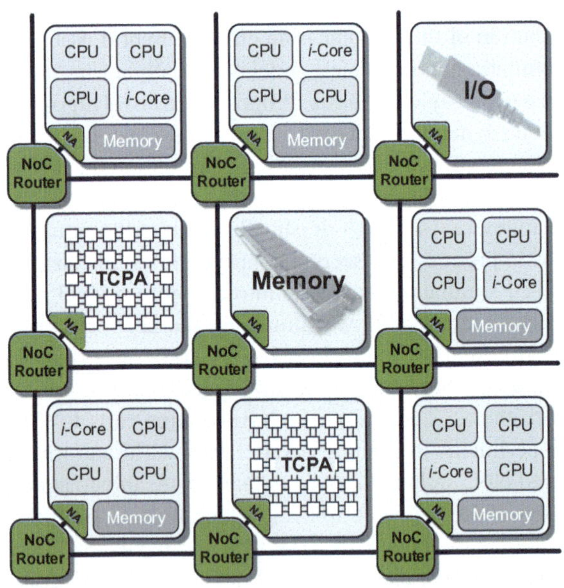

Fig. 4.1 The example architecture shows a heterogeneous tiled architecture (adapted from [3])

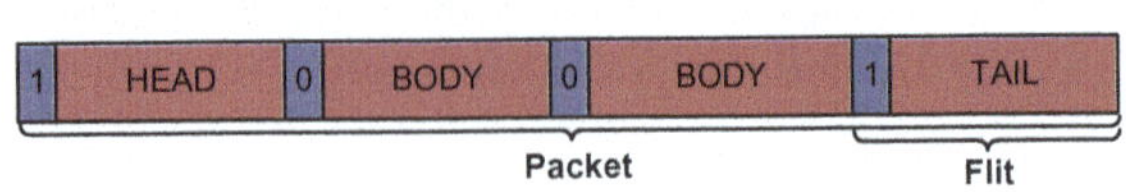

Fig. 4.2 NoC packet divided into header, body, and tail flits [4]

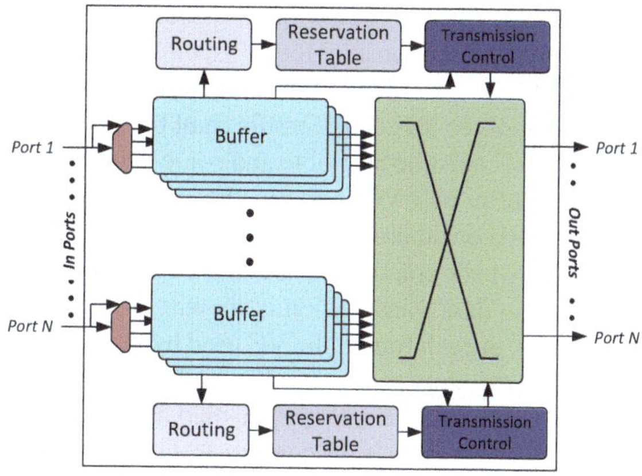

Fig. 4.3 Network router block diagram with virtual channels, routing logic, reservation table, transmission control unit, and crossbar [4]

The internals of a router are depicted in Fig. 4.3. A router is composed of several ports, one for each connected neighbor, components for routing, reservation, and transmission control and a crossbar, which connects the ports. Each port has a sending (Tx) and a receiving part (Rx). To increase the throughput of the NoC, so-called VCs are used, which are state-of-the-art to multiplex physical links [5]. The Rx-part of a port contains a separate buffer for each VC. Incoming flits of the same packet are directed to the same input buffer. This is realized by dedicated wires between the routers, which select the respective VC buffer, which was previously reserved by the header flit. To prevent the buffers from overflow, a credit-based flow control is used. If a header flit arrives, the router decides depending on the destination address and the routing algorithm to which output port the flit and its following ones are directed. In mesh-based NoCs, typically deterministic and deadlock-free *dimension-order routing* algorithms are used. However, our proposed technique can be applied also to adaptive routing schemes. Without loss of generality, XY-routing [6], a simple, deterministic and widely used routing algorithm, is assumed in the following. Here, a flit is sent first in x-direction until the destination x-coordinate matches the router's x-coordinate, then the flit is sent in y-direction until the final destination is reached. Once the routing decision is taken, a virtual channel and the respective input buffer in the next router on the path is reserved. This is achieved by adding an entry to the router reservation table. The flits of a VC input buffer are then transmitted, once the transmission control unit selects the associated entry in the reservation table.

Besides the transmission of data in a BE manner, some applications require a certain QoS through GS data transfer. This QoS can be achieved by allocating the different VCs appropriately, using strategies such as synchronous *time division multiple access* (TDMA), priority-based, or round-robin scheduling. While TDMA may lead to fragmentation and priority-based scheduling to starvation of messages with low priority, we will focus on a *weighted round-robin* (WRR) scheme as proposed in [4]. Here, it is possible to assign VCs to local TSs. Typically, there exist more Tss in a scheduling period than VCs per port, and through a *service level* (SL), the number of TSs that is assigned to a certain VC can be controlled. In contrast to synchronous TDMA, the position of the assigned TS can vary and is not globally fixed. For example, consider four VSs, three SLs and a scheduling period of six cycles (i.e., six TSs). Then, the first SL correspond to one TS, the second SL to two TSs, and SL 3 to three TSs. If BE traffic shall be used concurrently to GS connections, at least one VC must be available and some TSs should be reserved for the BE traffic. In case a buffer of a VC is empty, the corresponding TS can be used for scheduling flits from BE or other GS connections, which is an important difference compared to TDMA where empty slots are not used for other flits. In Fig. 4.4, an example of a concurrent data transmission is shown: *C1* is an established GS connection using SL 3. Thus, in each router, the VC used by the connection is assigned to 3 time slots. Different time slots can be used in each router. *C2* is a GS connection using SL 1. This connection is currently established, thus the header flit is shown. *C3* represents a BE data transmission. In contrast to GS connections, header, body (not shown), and tail flit are transmitted contiguously.

According to [4], upper bounds for throughput and latency can be calculated and guaranteed. For this purpose, the maximum number of utilized TSs on a GS connection path is defined as follows [4]:

$$TS_{util,max} = \max_{i \in [1,k]} \{TS_{util,i}\} \tag{4.1}$$

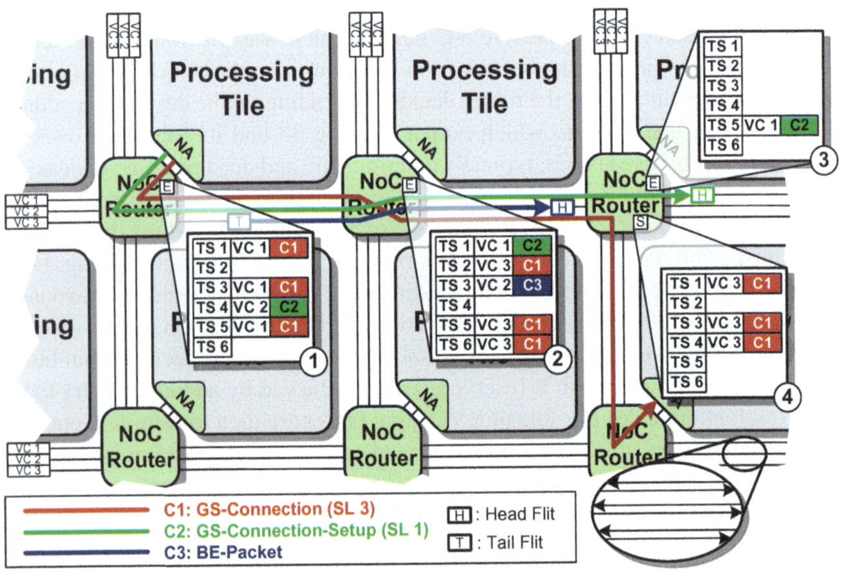

Fig. 4.4 Example communication scenario on considered tiled architecture. Three transmissions of different service levels are simultaneously active. A guaranteed service (GS) connection with a service level of 3 (C1), another GS connection with service level 1 (C2), and a best effort (BE) connection (C3) [4]

$TS_{util,i}$ denotes the used time slots at port i, which is on the GS path with the ports 1 to k. Together with the service level SL (ranging from 1 to SL_{max}) and the physical link bandwidth TP_{link}, the throughput TP_{GS} is given by [4]:

$$TP_{GS} = \frac{SL}{TS_{util,max}} TP_{link} \tag{4.2}$$

The worst case delay t_{GS} can be calculated as follows:

$$\begin{aligned} t_{GS} = S_{pkt} + \left\lceil \frac{S_{pkt}}{SL} \right\rceil (TS_{util,max} - SL) + t_{hop} \\ + (H-1)(TS_{util,max} - SL + t_{hop}) \end{aligned} \tag{4.3}$$

H denotes the length of the GS route in hops, t_{hop} is the delay per flit per hop, and S_{pkt} is the packet size in flits.

4.2 Hybrid NoC Simulation

As explained already before, a highly timing-accurate simulation of the behavior of a NoC including the delays of packet transmissions can be achieved by cycle-accurate simulation. In our system-level simulation, we are also very interested in such NoC delays, because they may highly influence the performance of distributed applications. Yet, the simulation overhead of cycle-accurate simulation of flit-by-flit data transport per each link would not be feasible. To this end, we propose a hybrid NoC simulation approach in order to speed up the time-consuming cycle-accurate NoC simulation while at the same time introducing only minor timing inaccuracies, which is explained in the following.

The main idea of our hybrid NoC simulation approach is to detect phases of predictable communication behavior during the simulation of packet transmissions through the NoC, and to calculate analytically latencies for these phases. Using these calculations, the simulation time as well as the number of sent and received flits for each transmission are forwarded. Hence, our hybrid NoC simulation approach is realized by alternating between phases of *cycle-accurate* simulation and *forwarding*. More specifically, we propose to simulate the phases of creation and the termination of transmissions in cycle-accurate mode and the transmission phase in a fast-forwarding mode. This decision is motivated by the fact that during creation and termination of transmissions, dynamic effects depending on the current router states such as reservation delays can happen. Timing predictions in this phase are difficult and would require complex data structures and calculations, which would diminish or even override the gain in simulation performance through forwarding, as will be shown.

4.2.1 Transmission Graph

We introduce a so-called *transmission graph* as main data structure used in our hybrid NoC simulation approach. This graph is used to determine the proportionate bandwidth values for all transmissions being currently active on the NoC depending on their reserved number of time slots and their congestion at physical NoC links at the same time. When the NoC enters a steady state, which means no creation or termination of any transmission is happening, these bandwidth values are then used to determine estimations on the latency of each transmission.

Definition 4.1 (*Transmission Graph* [1]) Let G be a directed bipartite graph $G = (V, E)$ where the node set V is partitioned into a set of link nodes V_L and transmission nodes V_T such that $V = V_L \cup V_T$, $V_L \cap V_T = \emptyset$. Edges can only exist between nodes from V_T to V_L, i.e., $E \subseteq V_T \times V_L$.

In particular, the two node types have the following meanings: Transmission nodes represent the proper transmissions. More formally, a transmission u is defined by a path through the NoC with respective source and destination coordinates, the type of traffic (BE or GS), the number of reserved time slots $ts(u)$ as well as the number of *flits*(u) to be transmitted. Link nodes represent the physical links in the NoC, which may be shared by multiple transmissions through the reservation of time slots at the sending part of the corresponding router port.

Note that each router port is divided into a sending and a receiving part and thus, has two physical links, one for outgoing and one for incoming flits. Each link node v has a certain number of time slots $ts(v)$, which is the sum of the time slots of all transmission nodes connected to it: $ts(v) = \sum_{u \in V_T:(u,v) \in E} ts(u)$. All transmission nodes that are connected to a link node share the available bandwidth proportional to their amount of reserved time slots. A transmission node can be connected to

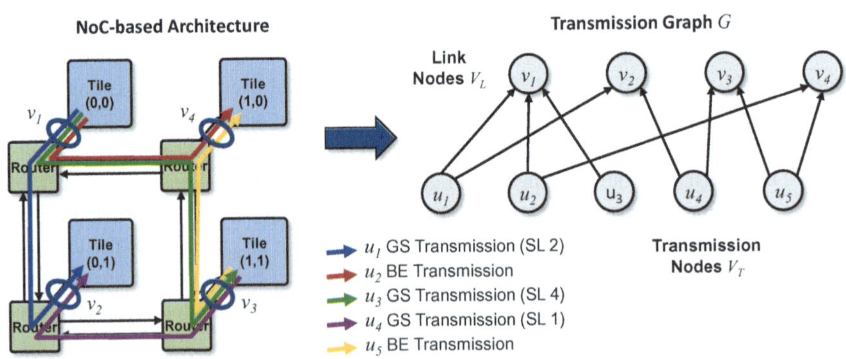

Fig. 4.5 On the left side, a snapshot in time of 5 active transmissions on a NoC-based architecture is shown. Each transmission sends flits along its path through the NoC from its source tile to its destination tile. On the right side, the transmission graph reflecting the overlappings of the transmissions is depicted [1]

Algorithm 4.1 Adding a new transmission u_{new} to the transmission graph $G = (V, E)$ with $V = V_L \cup V_T$.

```
procedure ADD-TRANSMISSION(G, u_new)
  V'_T := ∅
  V'_L := V_L
  for all u ∈ V_T do
    if IS-OVERLAPPING(u_new, u) then
      V'_T := V'_T ∪ {u}
    else
      V'_L := V'_L \ {v ∈ V_L : (u, v) ∈ E}
  for all v ∈ V'_L do
    V'_T := V'_T \ {u ∈ V_T : (u, v) ∈ E}
    E := E ∪ {(u_new, v)}
    ts(v) := ts(v) + ts(u_new)
  V_P := ∅
  for all u ∈ V'_T do
    if ∃P ∈ V_P : ∀u' ∈ P : IS-OVERLAPPING(u', u) then
      P := P ∪ {u}
    else
      V_P := V_P ∪ {{u}}
  for all P ∈ V_P do
    v_new = CREATE-LINK-NODE
    for all u ∈ P do
      E := E ∪ {(u, v_new)}
      ts(v_new) := ts(v_new) + ts(u)
    E := E ∪ {(u_new, v_new)}
    ts(v_new) := ts(v_new) + ts(u_new)
    V_L := V_L ∪ {v_new}
  V_T := V_T ∪ {u_new}
```

several link nodes as a transmission usually involves multiple hops and therefore multiple potentially shared links on the NoC. An example of a transmission graph for a snapshot in time of a simple communication scenario is shown in Fig. 4.5.

4.2.2 *Construction of the Transmission Graph*

Obviously, the transmission graph is a very dynamic data structure that must be constructed, respectively updated at run time, either once a new transmission is registered which depends on the communication behavior of the applications or once an active transmission has been terminated. Accordingly, new transmission nodes are added to or removed from the transmission graph. The algorithms explaining these operations in detail are shown in Algorithms 4.1 and 4.2.

In the following, we shortly summarize these two algorithms. If a new transmission node u_{new} has to be added to the transmission graph, it has to be checked first to which link nodes V'_L it needs to be connected. This is the case if the routing path of u_{new} passes through the considered link node. Here, the function IS-OVERLAPPING

Algorithm 4.2 Removing a transmission u_{del} from the transmission graph $G = (V, E)$ with $V = V_L \cup V_T$.

```
procedure REMOVE-TRANSMISSION(G, u_del)
    for all v ∈ V_L : (u_del, v) ∈ E do
        E := E \ {(u_del, v)}
        ts(v) := ts(v) − ts(u_del)
        if |{u ∈ V_T : (u, v) ∈ E}| = 1 then
            E := E \ {(u, v)}
            V_L := V_L \ {v}
    V_T := V_T \ {u_del}
```

(line 5 in Algorithm 4.1) is used, which determines whether the paths of two transmissions are overlapping in the sense that they share at least one physical NoC link. If new edges are created, at the same time, the time slots $ts(v)$ of the link nodes have to be updated. Second, the remaining overlapping transmission nodes V_T' have to be considered. This set is partitioned into subsets P of transmission nodes overlapping among each other. For each partition, a new link node v_{new} is created and all corresponding transmission nodes are connected to it. Removal of a transmission node requires to remove all edges originating from u_{del} to link nodes as well as to update their corresponding numbers of time slots $ts(v)$. If only one transmission node remains after removing the edges, the corresponding link node and the edge to it is also removed. The worst-case complexity of both algorithms for adding a new transmission to G as well as the removal is $\mathcal{O}(|V_T|^2)$.

4.2.3 *Update of the Transmission Graph*

After the transmission graph has been constructed, it can be used to determine the proportionate bandwidth values $bw(u)$ for each active transmission $u \in V_T$. These values depend on the number of reserved time slots $ts(u)$ as well as on the overlappings with other transmissions captured by the transmission graph. A detailed algorithm explaining how the bandwidth values $bw(u)$ are calculated, is shown in Algorithm 4.3. A quick summary on this algorithm is provided as follows: The algorithm assigns two temporary values to each link node $v \in V_L$. First, the remaining number of time slots $ts'(v)$ and second, the remaining bandwidth $bw'(v)$. These values as well as the bandwidth values of each transmission node are initialized at the beginning of the algorithm. The algorithm then computes the bandwidth values of each transmission node by iterating over the link nodes in ascending order according to the ratio $bw'(v)/ts'(v)$. This guarantees an assignment of the smallest possible bandwidth $bw(u)$ of the transmission u according to the following equation [1]:

$$bw(u) = \frac{bw'(v) \cdot ts(u)}{ts'(v)}$$

Algorithm 4.3 Algorithm to determine the proportionate bandwidth values $bw(u)$ of each transmission $u \in V_T$.

```
procedure UPDATE-TRANSMISSION-GRAPH(G)
    for all v ∈ V_L do
        ts'(v) := ts(v)
        bw'(v) := 1
    for all u ∈ V_T do
        bw(u) := 0
    V'_L := V_L
    while V'_L ≠ ∅ do
        v_min := argmin_{v∈V'_L} {bw'(v)/ts'(v)}
        V'_L := V'_L \ {v_min}
        V'_T := {u ∈ V_T : (u, v_min) ∈ E}
        while V'_T ≠ ∅ do
            u_max := argmax_{u∈V'_T} {ts(u)}
            V'_T := V'_T \ {u_max}
            if bw(u_max) = 0 then
                bw(u_max) := bw'(v_min) · ts(u_max)/ts'(v_min)
                if bw(u_max) > bw_max then
                    bw(u_max) := bw_max
                for all v ∈ V_L : (u_max, v) ∈ E do
                    ts'(v) := ts'(v) − ts(u_max)
                    bw'(v) := bw'(v) − bw(u_max)
    for all u ∈ V_T : bw(u) = 0 do
        bw(u) := bw_max
```

If the calculated bandwidth is higher than the maximum possible bandwidth bw_{max} determined by the size of the FIFO buffers in the routers, it is reduced to bw_{max}, which is given by the following equation:

$$bw_{max} = \min\left\{\frac{buffer\_size}{send\_delay + 1}, 1\right\}$$

For each processed transmission node, the number of remaining time slots $ts'(v)$ and the remaining bandwidth $bw'(v)$ of all connected link nodes are decreased. If a transmission is not overlapping with any other transmission, it gets assigned the maximum amount of bandwidth bw_{max}. The complexity of the bandwidth calculation in Algorithm 4.3 is also $\mathcal{O}(|V_T|^2)$.

4.2.4 Simulation Main Loop

The simulation main loop handles the coordination and transition between phases of cycle-accurate NoC simulation and fast-forwarding. Because our NoC simulator is integrated as part of a system-level simulator, the surrounding simulation framework

Algorithm 4.4 Main loop of the hybrid NoC simulation.

```
1 procedure NoC-Simulation-Main-Loop(G, max_cycles)
2     for i := 0 to max_cycles − 1 do
3         if num_border_flits = 0 then
4             Update-Transmission-Graph(G)
5             cycles := Apply-Forwarding(G, max_cycles − i − 1)
6             i := i + cycles
7         Simulate-One-Cycle
```

Algorithm 4.5 Forwarding algorithm using the transmission graph G.

```
1 function Apply-Forwarding(G, max_cycles)
2     for all u ∈ V_T do
3         cycles := (flits(u) − flits_sent(u) − 1)/bw(u)
4         max_cycles := min{cycles, max_cycles}
5     for all u ∈ V_T do
6         flits := RoundToNextInteger(max_cycles · bw(u))
7         flits_sent(u) := flits_sent(u) + flits
8         flits_received(u) := flits_received(u) + flits
```

determines a maximum amount of cycles the NoC simulator can advance and after which it has to give back control (*max_cycles*). The algorithm depicted in Algorithm 4.4 shows the main loop of our NoC simulator that is called by the system-level simulator. Basically, it simulates cycle-by-cycle until a steady communication state has been detected. Then, the forwarding algorithm is applied. As explained earlier, this condition is fulfilled for intervals free of any creation or termination of transmissions in the NoC. By implication, this means that neither a first nor a last flit of a transmission is located in any router buffer (head and tail flit in case of BE, first and last body flit in case of GS). We call these *border flits* and count their occurrences in the routers by tracking the sent and received flits of the transmissions (*num_border_flits*). If this amount becomes zero, forwarding is immediately applied.

The first step is to update the transmission graph to determine the latest bandwidth values for all active transmissions. Afterwards, the forwarding algorithm as described in Algorithm 4.5 is called. Here, the smallest number of cycles is determined by which all active transmissions can be *safely* forwarded. Obviously, a fast-forwarding phase must end before any transmission sends its last flit, because this has to be simulated again in cycle-accurate mode. Due to the rounding operation in line 6 of Algorithm 4.5, an error is introduced at the calculation of the number of FLITs being forwarded for each transmission, which will be evaluated later. The second part of the algorithm is to increase the number of sent and received flits for each active transmission based on its current bandwidth and the number of forwarded cycles. In the following, we introduce the programming framework in which the proposed hybrid simulation approach has been integrated for purpose of subsequent evaluation.

4.3 X10 Communication and InvadeSIM Integration

As mentioned already, for implementing invasive parallel applications, the parallel programming language X10 is used. It follows the PGAS programming model and supports distributed memory architectures by providing convenient language constructs for communication and data distribution. In this section, we will inspect the underlying communication mechanism of X10 in more detail to understand, how our efficient NoC simulation approach can be integrated into the X10 runtime as well as into InvadeSIM.

4.3.1 X10 Communication—Language Level

The PGAS programming model introduces the concept of so-called *places*, which represent logical partitions of a global address space. Physically, each place represents a separate shared-memory location with own processing resources. In case of a heterogeneous tiled processor architecture, a place corresponds to a single tile. An X10 program is able to distribute data to these places for parallel computations. Data objects can be created on any place and activities can access these data objects without NoC delays, if they are executed on the same place where the object was created (local access). Accesses to data objects that reside on a different place as where the activity is executed (remote access) require a place shift to the respective place. This causes network communication, which is hidden from the programmer, but he is aware that remote accesses are possibly delayed through network latencies. An underlying communication library is responsible for the data transport between the places over the NoC.

The X10 language construct that causes network communication is `at (P) {S}`, which is also referred to as place shift. Here, the statements represented by `S` are encapsulated in a so-called closure and all data objects from the surrounding context that are accessed by the statements are copied into the closure and transmitted to the remote place given by `P`. On this place, a new activity is created that executes the code of the closure. The following example demonstrates the usage of inter-tile parallelism in X10. In line 1, an list object is created on the current place: `Place(0)`. In line 4 and line 7, two activities are asynchronously created using `async at` on other places: `Place(1)` and `Place(2)`. The barrier synchronization caused by `finish` in line 3 on the current place waits until the spawned activities are finished on the remote places. Both remote activities access the list object in line 5 and line 8. However, due to the place shift and the X10 deep copy semantics, they operate on different copies on their own place. They are not manipulating the original list object on the first place and at the end of the finish statement, the initial list object is as empty as at the beginning.

This is an important consequence for an X10 programmer, because he has to be aware of the location of every data object he is manipulating and to perform a

```
val list = new ArrayList[int]();
/* Creating activities on other tiles */
finish {
    async at (Place(1)) {
        list.add(1);
    }
    async at (Place(2)) {
        list.add(2);
    }
}
```

place shift to that location before manipulating. Otherwise one can work on copies—however, this might lead to results, which are not intended.

4.3.2 X10 Communication—Runtime Level

We now take a look into the mechanisms, how these X10 communication constructs on language level are implemented. This investigation allows us determining when and how much network traffic is generated at run time of an X10 program, which is the input to our network simulation. Each X10 language construct is replaced by the X10 compiler by a call to the X10 runtime. Table 4.1 shows the mappings of some of the language constructs to their respective runtime calls. The `runAt` runtime call realizes the communication among places and hence, generates the network traffic.

X10 maps each place in the program to a separate process on the system. These processes execute the same binary and use a native communication library to communicate among each other. The `runAt` call at X10-level will result in a C++ native call, which serializes the data that has to be transmitted to the remote place. This includes all data objects that are accessed by the closure. The compiler knows already which data objects are accessed and prepares proper serialization methods for each closure. After serialization, the X10RT layer is used to transmit the data to the remote place. This is a C++ library that implements the communication between the places. It is designed to be able to use different communication libraries, which provide the actual communication between the places, e.g., MPI, Sockets, or IBM-specific protocols such as PAMI or BGP. The X10RT library is separated into three layers (front, logical, and net layer), which fulfill different tasks. An overview of the X10 communication layers is depicted in Fig. 4.6.

In our system-level simulator InvadeSIM, we replaced the communication layers of the X10RT with our hybrid NoC simulator (see Fig. 4.6) by forwarding all communication requests including target place and payload to our NoC simulator, which then simulates the corresponding NoC delay times in sync with the processor simulation. Once a communication request finishes, an appropriate activity is created at the remote place, which operates on the received data. All communication requests are

Table 4.1 Mapping of X10 language constructs for concurrency, distribution, and synchronization to the X10 runtime

X10 language construct	X10 runtime call
`async`	`Runtime.runAsync()`
`at`	`Runtime.runAt()`
`finish`	`Runtime.startFinish()` `Runtime.stopFinish()`
`atomic`	`Runtime.enterAtomic()` `Runtime.exitAtomic()`

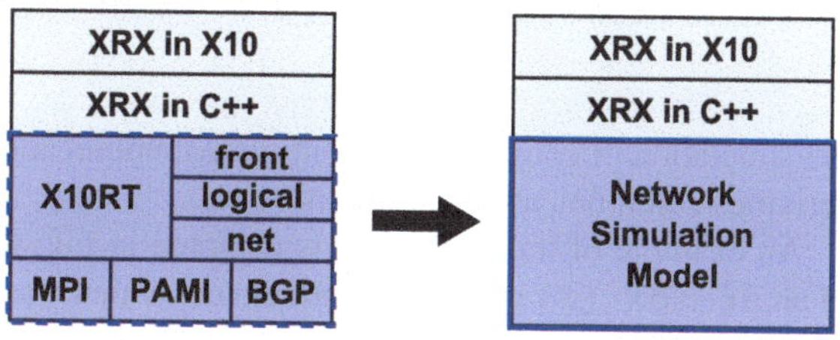

Fig. 4.6 The different layers of the standard X10 communication primitives between places for the C++ back end and their replacement by our network simulation model

non-blocking, which enables asynchronous communication with remote places. In order to wait for the result of a remote computation, the respective activity is blocked until it is deblocked by an activity, which is sent back by the remote place after the computation has finished.

4.3.3 InvadeSIM Integration

Another important task was the integration of the NoC simulation into the existing simulation framework. This framework called InvadeSIM for high-level timed functional MPSoC simulation was introduced earlier. Here, multiple parallel applications executing on a heterogeneous architecture can be simulated. Timing estimations for different processor types and a synchronization mechanism to execute several simulation tasks by a simulation thread are the key concepts of this framework. Each simulation tasks maintains its own simulation clock and advances in different time steps. At a synchronization point, the current simulation task is stopped and the simulation task with the smallest simulation time is selected for being executed next. In order to integrate the NoC simulation, it has to be wrapped as hardware simulation task. It remains inactive as long as no communication request occurs. If an activity executes a communication operation, it becomes active, inherits the current simulation time of the activity, and it put onto the min-heap. Once selected, it begins simulating the communication request, either flit-by-flit or by forwarding until the time of the next event in the min-heap or the communication is done. This enables

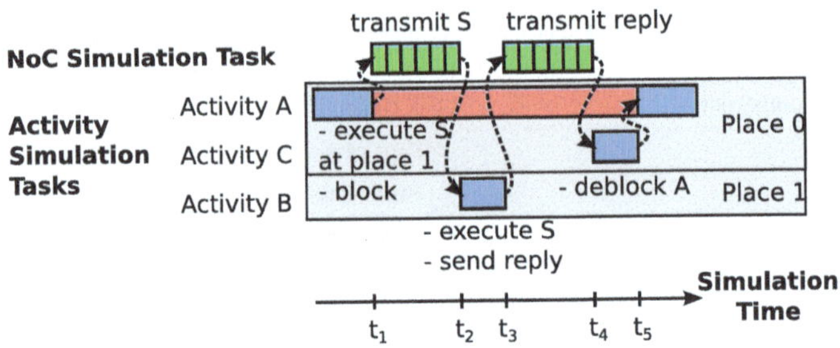

Fig. 4.7 Representation of the NoC simulation in a separate thread (green) and the synchronization with the activity simulation threads (blue—activity active, red—activity blocked) on the basis of an `at` communication pattern

the simulation of congestion of multiple communication requests caused by several activities communicating at the same time.

An example of this procedure is depicted in Fig. 4.7. Here, the communication of an `at (p) {S}` construct is simulated, which was initiated by activity A. First, activity B is created at place 1 and executes the set of statements `S`. During this time, activity A is blocked and waits for the result. After activity B is finished, a reply is sent back to place 0 in form of activity C, which then deblocks activity A. Each arrow in this diagram represents the handover between simulation tasks.

4.4 Experimental Results

By integration of our hybrid NoC simulation approach into a system-level simulator, which is able to simulate heterogeneous tile-based MPSoCs as shown in Figs. 4.4 and 4.5, we were able to test our approach for real-world parallel and distributed X10 applications. In the following, the introduced hybrid NoC simulation approach is evaluated in terms of accuracy and speedup in simulation latency compared to a cycle-accurate NoC simulation for a benchmark of distributed X10 applications, a complex streaming-based object-detection application as well as synthetically generated traffic. For the following experiments, we used an Intel Core i7 with a maximum clock frequency of 3.6 GHz and 8 MiB of RAM as host machine.

4.4.1 Parallel Applications

For the experiments, X10 applications from the IMSuite [7] benchmark are used, which contains 12 essential algorithms on graphs covering important and complex problems in the domain of distributed systems such as breadth first search, routing table creation, dominating set, spanning tree, vertex coloring, etc. These algorithms distribute their input data among all tiles of the target architecture, start computation, communicate and synchronize in between, and finally gather the partial results from the tiles and merge them into a final result. IMSuite allows generating different input data for the algorithms, e.g., in terms of graph size (-sz) and graph characteristics. For each algorithm, we created an input set with sizes ranging from 8 to 512 and the characteristics of a sparse maximum graph (-spmax). We simulated these applications on differently sized NoC architectures in order to investigate, how scalable our hybrid NoC simulation is. Our considered target architecture dimensions are: 4×4, 8×8, 16×16, and 32×32. Each tile has one processor core, i.e., for the 32×32 tiled architecture $1,024$ cores are simulated using our system-level simulator. Here, the functionality as well as the timing of the application threads on the cores are simulated in a processor simulation phase until a dynamic communication request initiates a NoC simulation phase (i.e., the system-level simulation executes the algorithm ADD- TRANSMISSION in Algorithm 4.1 and gives the control to the NoC simulation by calling the algorithm NOC- SIMULATION- MAIN- LOOP in Algorithm 4.4). A communication request contains the source and destination tile as well as the amount of bytes that have to be transmitted. For comparison of timing accuracy, the *latency* of a transmission is defined as the difference between the time the communication was requested at the source tile and the time the last flit has been received at the destination tile.

The first evaluation considers the accuracy of our simulation approach. Here, we determined the simulated latency of all transmissions an application caused during its execution for both simulation techniques and calculated the relative error of our hybrid simulation approach with respect to cycle-accurate simulation. The results of this evaluation for two different input sizes are depicted in Figs. 4.8 and 4.9. They

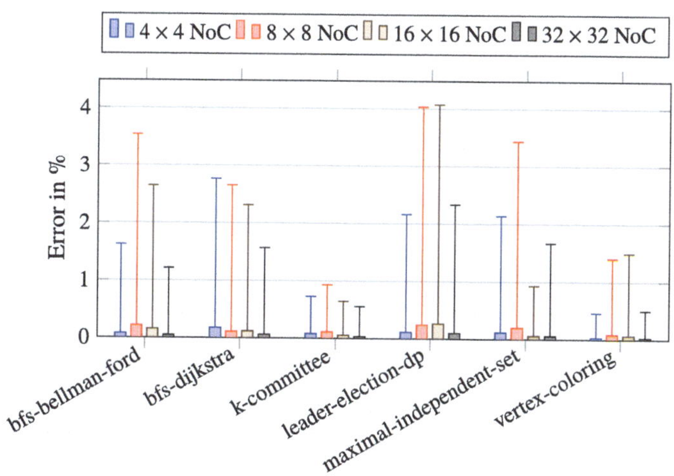

Fig. 4.8 Relative error of the simulated transmission latencies (averaged over all transmissions) introduced by the hybrid NoC simulation approach compared to cycle-accurate simulation for small problem inputs (-sz 16)

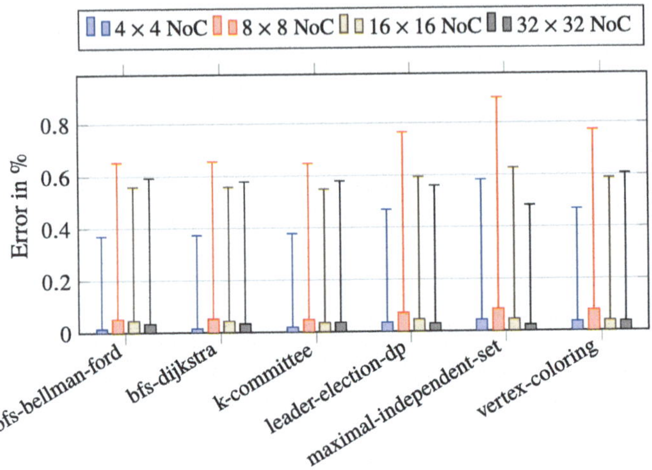

Fig. 4.9 Relative error of the simulated transmission latencies (averaged over all transmissions) introduced by the hybrid NoC simulation approach compared to cycle-accurate simulation for large problem inputs (-sz 512)

show the relative error averaged over all transmissions as well as confidence intervals with a confidence level of 95%. First, it is mentionable that for all considered benchmark programs, the average error amounts to less than 1%, which is extremely accurate and sufficiently accurate for many purposes and applications. However, the confidence intervals show variations of up to 4%, which indicates that the latency error for a couple of transmissions may be slightly higher than the average. This results from error propagation, because thousands of transmissions utilize the capacity of the NoC. Sometimes they are delayed until a used NoC channel becomes available. Here, errors coming from the forwarding approximation are propagated and differ a lot depending on the NoC and input size. Moreover, it can be seen that the error as well as its variation is much smaller in case of applications of large input sizes (see Fig. 4.9). This can be explained by the fact that the algorithms send smaller packets in case of smaller input sizes, see also Table 4.2. In effect, the error propagation applies much quicker than in case of large input sizes. Furthermore, the higher the latency of a transmission, the smaller the relative error can become in case of similar absolute errors.

The second evaluation considers the gain in simulation performance of our approach. Here, we determined the total time the simulator spends in NoC simulation for both simulation techniques and calculated the speedup of our hybrid simulation approach with respect to cycle-accurate simulation. Again, we considered two different input sizes. The results of this evaluation are depicted in Figs. 4.10 and 4.11. A first observation reveals that we are able to be more than 1,500 times faster than a cycle-accurate simulation for sufficiently large input data sizes (see Fig. 4.11). Furthermore, it can be observed that the simulation of small NoC architectures (e.g., 4 × 4) can be better accelerated than large ones. The larger the NoC

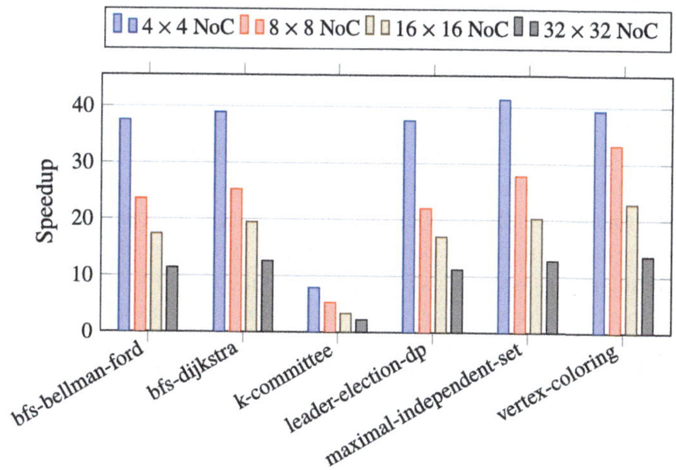

Fig. 4.10 Speedup of the hybrid NoC simulation approach compared to cycle-accurate NoC simulation for small problem sizes of input data (-sz 64)

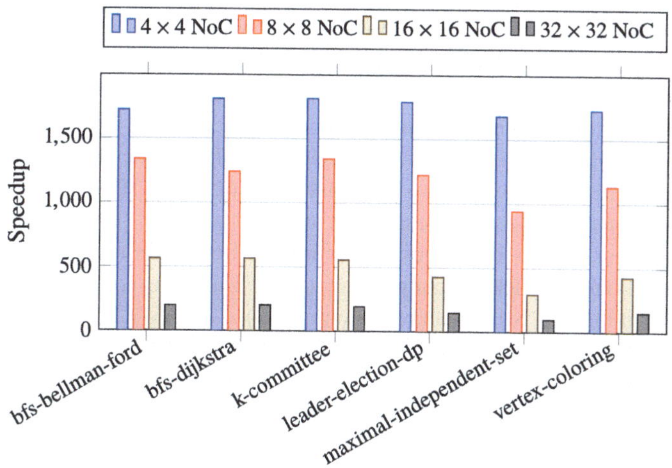

Fig. 4.11 Speedup of the hybrid NoC simulation approach compared to cycle-accurate NoC simulation for large problem sizes of input data (-sz 512)

becomes, the longer are the communication paths through the NoC and therefore, the cycle-accurate phases in case of reservations and terminations of transmissions, which finally leads to smaller observable speedup factors. However, it has to be mentioned that in the considered benchmark scenarios, the applications are distributed among the whole architecture, whereas real-world distributions would probably follow a cluster-based approach, which in turn would result in smaller communication paths and therefore in even higher speedup factors. Finally, the speedup increases for larger problem instances, as the more data has to be transmitted the more time can be saved by the fast-forwarding approach.

In Table 4.2, results for the bfs-dijkstra application on a 16 × 16 architecture are depicted in order to show the dependence of achievable speedup on the input size (-sz). The input data size is varied from 8 to 512, which results in a different amount of transmissions to be simulated and different average packet sizes in terms of flits. In order to run a sufficiently number of test runs for each scenario, we limited the simulation time to 5 minutes of each run. Hence, as can be seen in the table, the number of simulated transmissions for input sizes of 256 and 512 is smaller than for 64 and 128. Note that otherwise, a cycle-accurate NoC simulation would require days to simulate a full application scenario, e.g., on a 32 × 32 architecture with billions of flits to be transmitted. Instead, we stopped the system simulation with cycle-accurate NoC simulation after a timeout of 5 minutes and used the number of simulated NoC cycles to stop the system simulation in case of fast-forwarding. This allowed us to generate results for more than 300 application executions in an adequate time effort. The maximum simulation performance in terms of NoC cycles/s for the bfs-dijkstra application is 144M cycles/s, when fast-forwarding is used and 254k cycles/s in case of cycle-accurate simulation. Among all tested application scenarios, we achieved a maximum simulation performance of up to 683M cycles/s. For comparison, typical SystemC-based cycle-accurate NoC simulators are able to simulate up to 10k cycles/s and C++-based simulators up to 100k cycles/s [8]. Whereas, our own cycle-accurate NoC simulator achieves a maximum of 383k cycles/s.

In order to show the advantages of our NoC simulation for data-intensive real-world multi-media applications, we evaluated a distributed X10 version of a SIFT-based object-detection algorithm task chain as proposed in [9]. The goal of this task chain is to detect objects in images. It is composed of seven tasks: (1) An image source task, (2) an edge detection algorithm, (3) the Harris corner detector, (4) SIFT feature description, (5) SIFT feature matching, (7) a RANSAC-based registration, and (6) an image sink task. Each task is mapped on another tile on the simulated 3 × 3 target architecture and they communicate via the NoC. Experiments with full HD images (1920 × 1080 pixels) show a speedup of our hybrid NoC simulation compared to cycle-accurate NoC simulation of 6,200, while preserving an accuracy in terms of average transmission latency of 0.1457% with a 95-confidence interval of ±2.7083%.

Finally, we determined, how much the system simulator including CPUs and NoC can benefit from our proposed hybrid NoC simulation. Therefore, we measured the total time required for the entire NoC and functional simulation of an application execution for both techniques. The speedup numbers for the bfs-dijkstra application can be seen in the last column of Table 4.2.

Table 4.2 Simulation and performance results of the bfs-dijkstra application on a 16 × 16 architecture for different input sizes

Input size (-sz)	Simulated transmissions	Average packet size [flits]	NoC simulation performance [cycles/s]		Speedup system sim
			Cycle-accurate	Fast-forwarding	
8	3215	20.3	247 k	283 k	1.1
16	4131	56.5	253 k	479 k	1.8
32	5817	238.4	276 k	1521 k	4.9
64	13003	1155.8	248 k	4825 k	16.9
128	10761	5207.2	215 k	11423 k	46.9
256	3654	12754.5	229 k	56496 k	205.2
512	1632	32784.9	254 k	144132 k	408.7

4.4.2 Generated Traffic

Our last experiments investigate the margins of our NoC simulation approach in terms of accuracy and performance with respect to synthetically generated network traffic characterized by injection rate and packet size. For this, we simulated generated network traffic according to the *bit reversal* traffic pattern and again measured the relative latency error and speedup compared to cycle-accurate simulation. The results are shown in Figs. 4.12 and 4.13. It can be seen that the error is varying with an increased injection rate in case of around 100 flits packet size, but still below 4% and completely zero for packet sizes less than 10 flits. As for the speedup, it can be seen that it is almost independent of the injection rate, but in fact increases linearly with the packet size. For very small packets, the performance gain is almost zero. In such a scenario, too many short transmissions are alternating and only the cycle-accurate simulation mode is active, forwarding cannot be applied. Here, it is important to

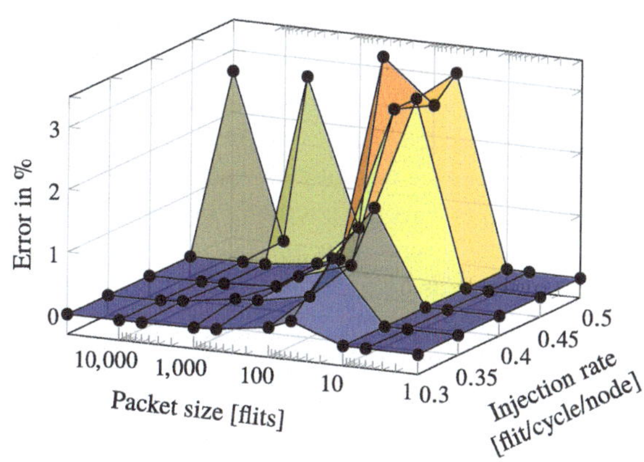

Fig. 4.12 Relative error of the simulated transmission latencies (averaged over all transmissions) introduced by the hybrid NoC simulation approach compared to cycle-accurate simulation for different packet sizes and injection rates of the synthetic traffic pattern *bit reversal*

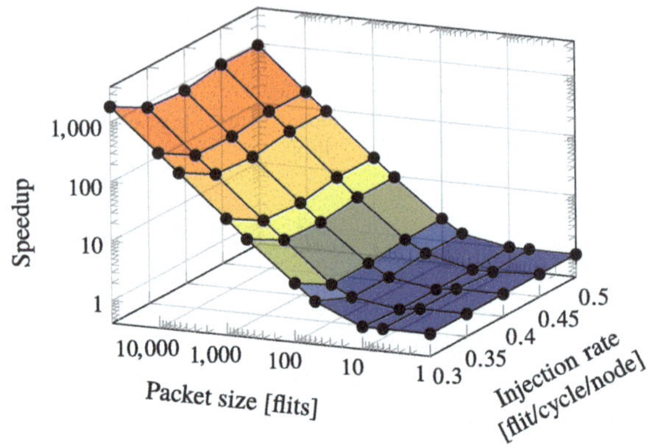

Fig. 4.13 Speedup of the hybrid NoC simulation approach compared to cycle-accurate NoC simulation for different packet sizes and injection rates of the synthetic traffic pattern *bit reversal*

mention that our approach is in worst-case as slow as cycle-accurate simulation, but provides a huge improvement in performance for real-world cases.

4.5 Related Work

There exist several different types of NoC simulators, which are briefly summarized in the following. Most accurate NoC simulators are based on RTL simulations, where a model of the NoC is written in a hardware description language and synthesized into hardware. These are by far the slowest ones.

Faster NoC simulators are written in C/C++ or SystemC and can be driven by synthetic traffic generators, recorded traffic traces, or in-the-loop real application traffic. Detailed cycle-level pure NoC simulators are, e.g., Noxim [10], BookSim [8], DARSIM [11], OMNeT++ [12], and Nirgam [13]. In the Nirgam simulator, mesh and torus topologies of NoCs can be evaluated in terms of latency and throughput. Synthetic traffic generators for a *constant bit rate (CBR)* or *bursty* traffic are available. In the Noxim simulator, the user can choose between Poisson, burst, pseudo-Pareto, or custom distribution of network traffic. Also, it is possible to define hotspot nodes, which receive data with a higher probability. In [14], a NoC simulation platform based on the widely used C++ network simulation framework OMNeT++ is presented. The authors extended OMNeT++ by NoC requirements and developed a tool, which provides an interface to a language for describing topologies, offers a rich graphical user interface, and works internally highly parallel. The user can configure single nodes in order to send data to a certain or random destination. Another option is to use trace files as input for the traffic generation. Without additional tools, these NoC simulators are mainly driven by traffic generators or traces based on workload descriptions such as task graphs like presented in [15].

Yet, neither these task graphs, nor simple traffic generators describe the real behaviour of packet traffic that is generated by parallel programs executed on a

NoC-based MPSoC platform. In contrast, our NoC simulator is an integrated part of a full-system simulator and can be driven by dynamic real parallel application traffic. Other full-system simulators for multi-core architectures including NoC simulation are GARNET [16] and HORNET [17]. They provide highly configurable NoC models and apply cycle-level NoC simulation. HORNET and [18] apply also techniques for speeding up NoC simulation by parallel simulation. Our acceleration technique for NoC simulation is based on forwarding. It provides much higher speedups of up to three orders of magnitude with respect to cycle-accurate simulators, while keeping the accuracy of above 95%. Further techniques for speeding up NoC simulation are sampling-based approaches, which are investigated, e.g., in [19] for thread-parallel C/C++ applications. Common to our following hybrid approach, the NoC simulation process is divided into detailed and forwarding phases. However, a time-consuming parameter selection phase has to be performed. Our simulation approach can be applied without changes to any dynamic communication traffic that is produced during the simulation of the execution of parallel programs.

Finally, an analytical approach for NoC performance analysis is investigated in [20]. Here, the authors present a mathematical model for on-chip routers and analyze a NoC regarding different distributions of flit arrival times in order to generate network traffic. They are able to analyze a NoC configuration an order of magnitude faster than through simulation. However, this approach requires to analyze an application beforehand, instead of employing the inherently generated network traffic including timing of a parallel application obtained by system simulation like in our approach.

4.6 Summary

In this chapter, we presented a novel hybrid NoC simulation approach for the use in system simulators to accelerate the simulation of NoC traffic, which is crucial for performance evaluation of distributed applications as well as to identify any bottlenecks in an underlying architecture. The main idea of this approach is to detect steady communication intervals during the transmission of data, which are then exploited to forward the simulation time. We introduced a so-called transmission graph, which allows determining the individual bandwidth values for each transmission and forwarding the NoC simulation time accordingly in intervals between arrivals and terminations of transmission requests coming from the simulated applications. Furthermore, we presented an integration technique to incorporate this NoC simulation approach into InvadeSIM. Here, we investigated the underlying communication mechanism of X10 to extract the network traffic from X10 programs to feed the NoC simulator. Experiments using distributed applications from the IMSuite benchmark showed an average simulated latency error of less than 5%, while accelerating the simulation time for the NoC simulation by a factor of 1,000 and many cases.

References

1. S. Roloff, F. Hannig, J. Teich, High performance network on-chip simulation by interval-based timing predictions, in *Proceedings of the 15th IEEE/ACM Symposium on Embedded Systems for Real-Time Multimedia (ESTIMedia)*, Seoul, Republic of Korea, 15–20 October 2017 (ACM, 2017), pp. 2–11. ISBN: 978-1-4503-5117-1. https://doi.org/10.1145/3139315.3139320.
2. S. Roloff et al., NoC simulation in heterogeneous architectures for PGAS programming model, in *Proceedings of the 16th International Workshop on Software and Compilers for Embedded Systems (M-SCOPES)*, St. Goar, Germany, 19–21 June 2013 (ACM, 2013), pp. 77–85. ISBN: 978-1-4503-2142-6. https://doi.org/10.1145/2463596.2463606.
3. J. Heisswolf et al., The invasive network on chip-a multi-objective many-core communication infrastructure, in *Workshop Proceedings of the 27th International Conference on Architecture of Computing Systems (ARCS)*, Lübeck, Germany, 25–28 Feb 2014 (VDE Verlag, 2014), pp. 1–8. ISBN: 978-3-8007-3579-2
4. J. Heisswolf, R. König, J. Becker, A scalable NoC router design providing QoS support using weighted round robin scheduling, in *IEEE 10th International Symposium on Parallel and Distributed Processing with Applications (ISPA)*, Leganes, Spain, 10–13 July 2012 (IEEE, 2012), pp. 625-632. ISBN: 978-1-4673-1631-6. https://doi.org/10.1109/ISPA.2012.93.
5. W.J. Dally, Virtual-channel flowcontrol. IEEE Trans. Parallel Distrib. Syst. **3**(2), 194–205 (1992)
6. L.M. Ni, P.K. McKinley, A survey of wormhole routing techniques in direct networks. Computer **26**(2), 62–76 (1993). ISSN: 0018-9162
7. S. Gupta, V. K. Nandivada, IMSuite: a benchmark suite for simulating distributed algorithms, in *The Computing Research Repository (CoRR)* (2013). arXiv: 1310.2814 [cs.DC]
8. N. Jiang et al., A detailed and flexible cycle-accurate Network-on-Chip simulator, in *Proceedings of the International Symposium on Performance Analysis of Systems and Software*, April 2013, pp. 86–96
9. S. Roloff et al., ActorX10: an actor library for X10, in *Proceedings of the 6th ACM SIGPLAN Workshop on X10 (X10)*, Santa Barbara, CA, USA, 14 June 2016 (ACM, 2016), pp. 24–29. ISBN: 978-1-4503-4386-2. https://doi.org/10.1145/2931028.2931033.
10. V. Catania et al., Noxim: an open, extensible and cycle-accurate network on chip simulator, in *Proceedings of International Conference on Application-Specific Systems, Architectures and Processors*, 14 July 2015, pp. 162–163
11. M. Lis et al. DARSIM: a parallel cycle-level NoC simulator, in *Sixth Annual Workshop on Modeling, Benchmarking and Simulation (MoBS)*, June 2010
12. A. Varga, R. Hornig, An overview of the OMNeT++ simulation environment, in *Proceedings of the 1st International Conference on Simulation Tools and Techniques for Communications, Networks and Systems & Workshops 2008*, pp. 60:1–60:10
13. M. S. Gaur et al., NIRGAM: a simulator for NoC interconnect routing and applications' modeling. in *Tool Presentation at the University Booth at Design, Automation and Test in Europe (DATE)*, Acropolis, Nice, France, April 2007. https://www.date-conference.com/files/file/10-ubooth/ub-1.4-p04.pdf
14. Y. Ben-Itzhak et al., NoCs simulation framework for OMNeT++, in *Fifth ACM/IEEE International Symposium on Networks on Chip (NOCS)*, May 2011 (ACM, 2011), pp. 265–266. https://doi.org/10.1145/1999946.1999993.
15. J. M. Joseph, T. Pionteck, A cycle-accurate network-on-chip simulator with support for abstract task graph modeling, in *International Symposium on System-on-Chip (SoC)*, October 2014, pp. 1–6. https://doi.org/10.1109/ISSOC.2014.6972440.
16. N. Agarwal et al., GARNET: a detailed on-chip network model inside a full-system simulator, in *Proceedings of the International Symposium on Performance Analysis of Systems and Software*, April 2009, pp. 33–42
17. P. Ren et al., HORNET: a cycle-level multicore simulator. Trans. Comput.-Aided Design Integr. Circ. Syst. **31**(6), 890–903 (2012)

18. M. Eggenberger, M. Radetzki, Scalable parallel simulation of networks on chip, in *Proceedings of the International Symposium on Networks-on-Chip*, April 2013, pp. 1–8
19. W. Dai, N. E. Jerge, Sampling-based approaches to accelerate network-on-chip simulation, in *Eighth IEEE/ACM International Symposium on Networks-on-Chip (NoCS)* (IEEE, 2014), pp. 41–48. https://doi.org/10.1109/NOCS.2014.7008760.
20. Umit Y. Ogras, Paul Bogdan, Radu Marculescu, An analytical approach for Network-on-Chip performance analysis. IEEE Trans. Comput-Aided Design Integr. Circ. Syst. **29**(12), 2001–2013 (2010). https://doi.org/10.1109/TCAD.2010.2061613

Chapter 5
Parallel MPSoC Simulation and Architecture Evaluation

Innovative simulation mechanisms at system-level are key for embedded hardware designers and parallel software developers in order to predict the performance or other metrics of their particular designs or programs. This is important especially in a very early development phase where design space exploration techniques are applied to guide design decisions in proper directions or hardware is not available yet for software development and test. In case of modern parallel multi-core or many-core systems, which may contain hundreds to thousands of heterogeneous cores arranged in tiles with local memory, and connected via a *network-on-chip* (NoC), *design space exploration* (DSE) can be very costly and time consuming if simulation techniques are used for the exploration of non-functional properties such as performance. In order to provide performance estimations for parallel software and enable *architecture design space exploration*, fast functional and timing simulation techniques are required. Here, InvadeSIM was introduced in Chap. 3, which provides an efficient direct-execution simulation approach to tackle these goals. However, the presented simulation technique does not exploit the performance potential of modern multi-core host machines since the simulation is sequentialized. Thus, in this chapter, we present different parallelization techniques to accelerate the direct-execution simulation approach used in InvadeSIM. Four novel parallel discrete-event simulation techniques are proposed which map activity-level parallelism within the applications to core-level parallelism on the target architecture and back to thread-level parallelism on the host machine. In order to achieve this, the correct synchronization and activation of the host threads is necessary which constitutes a main focus of this chapter. Real-world parallel applications are used in experiments to compare the proposed techniques against each other and to demonstrate that a parallelized simulation can achieve speedups of up to 10 compared to a sequential simulation on a 12-core Intel Xeon processor. Since parallel simulation enables the evaluation of complex MPSoC architectures, we further present a case study about performance and costs trade-offs of different heterogeneous invasive architecture variants. This

S. Roloff et al., *Modeling and Simulation of Invasive Applications and Architectures*, Computer Architecture and Design Methodologies,
https://doi.org/10.1007/978-981-13-8387-8_5

includes design choices such as number of tiles, internal tile structure and selection of cores within each tile. The concepts for our parallel simulation and synchronization techniques have been published in [1] and the architecture evaluation case study in [2].

5.1 Parallel Simulation

Current simulators available in industry and academia range from highly detailed *cycle-accurate* to *timing-approximate* simulations. Cycle-accurate simulators may in general provide the best results in terms of timing accuracy but are mainly used for designing systems with a single or small number of cores. However, techniques at this level of detail may not be used to evaluate large-scale multi-core systems including hundreds or more cores because simulations would be too slow. During the last years, researchers have put a lot of effort to improve cycle-accurate simulation and to overcome its drawbacks, while keeping accuracy in an acceptable range. Two key problems were tackled: First, raising the abstraction level in multi-core simulation from cycle-accuracy to higher abstraction levels may noticeably improve simulation performance, but in general at the expense of accuracy in terms of accurate timing estimations for each event of the simulated target system. Second, parallel discrete-event simulation and proper synchronization techniques have been investigated to exploit the host cores of modern multi-core processors. These approaches include techniques for advancing simulated cores in lock-step manner for high accuracy but low simulation speed, or relaxed synchronization for high simulation speed but reduced accuracy.

Despite many achievements of recently proposed simulators that address fast and scalable multi-core simulation, they are mainly industry-driven and target only single, dedicated processor types or specific architectures. Such simulators often still model a lot of details, e.g., of the processor pipeline, branch prediction, and cache behaviors. However, micro-architectural details are often not required in many practical design studies at system-level and in parallel software development. With these deficiencies in mind, we propose parallel simulation concepts on the basis of the previously introduced simulation framework InvadeSIM as shown in Fig. 5.1. InvadeSIM uses a timing-approximate functional simulation approach (see Chap. 3), which tackles the complexity and the heterogeneity of modern multi-core architectures. It allows modeling different tiled MPSoC architectures and hence supports the DSE of architecture decisions such as number of tiles, NoC parameters, and number and type of cores on each tile. It is based on direct execution and a linear timing estimation model (time warping), which takes the computational properties of possibly different types of cores into account. It delivers important timing information about the parallel execution and competition of multiple applications with highly dynamic parallelism profiles and thus, enables the investigation of the computing paradigm invasive computing. However, the simulation approach introduced so far is completely sequentialized and does not exploit today's multi-core processors available

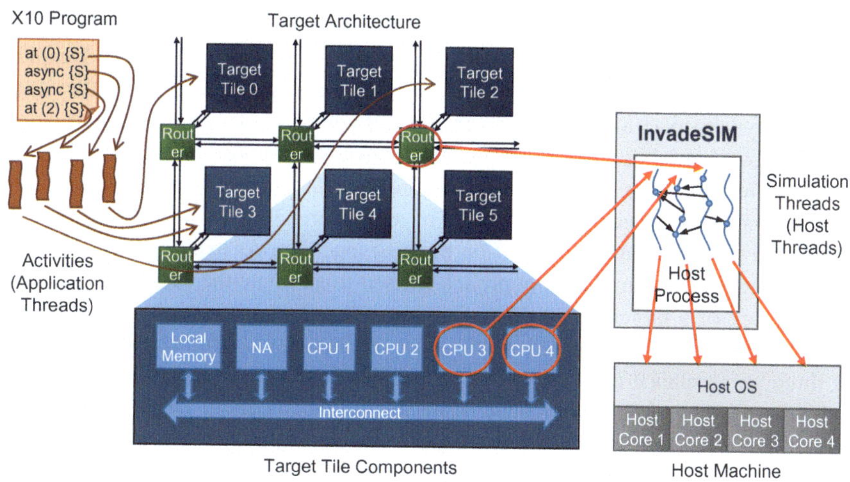

Fig. 5.1 Overview of our parallel MPSoC simulation approach and class of considered target architectures [1]

also inside a host machine. This prevents the investigation of realistic parallel applications and run-time system behavior, which require long running simulations.

Here, we propose and compare four novel parallelization and synchronization techniques to exploit the computing power of modern multi-core host machines (see also [1]). More specific, we combine the direct-execution simulation approach with these parallelization techniques by mapping activity-level parallelism inherently given in parallel X10 programs to core-level parallelism on the target architecture and back to thread-level parallelism on the host machine, while keeping the cores simulating threads in-sync. It is thereby shown that depending on the parallel workload and the proper synchronization technique, significant speedups of simulation performance may be achieved.

5.1.1 *Direct-Execution Processor Simulation*

The InvadeSIM simulation kernel simulates the functional execution of spawned activities on individual processors as well as their timing, which in general depends on the type of processor assigned to each activity and its frequency. As explained in Chap. 3, each simulation of an activity on a target core is wrapped by a software simulation task that is executed by a simulation thread (e.g., as indicated by a red arrow from CPU3 to a simulation thread of InvadeSIM in Fig. 5.1). Sequential simulation uses only a single simulation thread, while parallel simulation exploits more than one as depicted in Fig. 5.1. Using parallel discrete-event simulation, the simulation threads are then assigned by the host operating system for parallel execution to the cores of the host machine. However, the simulation threads have to be properly synchronized in order to satisfy global causality of all events (see Sect. 3.3.2). Despite there is a distinction between *hardware* and *software* simulation threads, they use the same synchronization primitives. In the following, the simulation of software threads is considered in more detail and some basic definitions are introduced.

The simulation of a single processor core $core(task(d))$ executing an activity $activity(task(d))$ mapped to it is wrapped by a software simulation task $task(d)$ that is executed by a simulation thread d on the host machine. As explained in Chap. 3, our direct-execution simulation approach is used to simulate the functional as well as the timing of a target core executing an activity. During functional simulation, the application code is executed on the host machine by a simulation thread and the three performance counter values c_0 (retired instructions), c_1 (L1 data cache read misses), and c_2 (L1 data cache write misses) are counted by a performance counter. Each simulation thread d has its private performance counter $pc(d)$. Each functional simulation phase is followed by a timing simulation phase, where the execution latency Δt on the target processor is predicted and added to the simulation time $time(task(d))$ of the current simulation task $task(d)$ of simulation thread d. Δt is determined by the already introduced *time warping* mechanism (see Sect. 3.3.1), which requires the three performance counter values (c_0, c_1, and c_2), the calibrated weights of the processor (p_0, p_1, and p_2), and the processor clock frequency f. The following list summarizes the introduced terms.

- d—Simulation thread
- $pc(d)$—Performance counter of simulation thread d
- $task(d)$—Current simulation task of simulation thread d
- $time(t)$—Simulation time of simulation task t
- $core(t)$—Processor core of software simulation task t
- $activity(t)$—Activity of software simulation task t

During functional simulation, there is no information available about how time has been advanced in the meanwhile and $time(task(d))$ remains the same until functional simulation is stopped. The duration of a functional simulation phase depends on the actual application code. For instance, if an operation is executed that manipulates the underlying hardware model, e.g., creating a new activity on another processor core, timing synchronization with all other simulation threads is required. These events are called *synchronization points* and have been introduced in Sect. 3.3.2. In a sequential simulation, only one simulation thread pops simulation tasks from the global min-heap. If an activity of a software simulation task executes a blocking operation, its context is saved and it is removed from execution. The simulation thread pops the next simulation task from the min-heap and switches to its context to continue its execution. In a parallel simulation, more simulation threads are available for executing simulation tasks. However, the problem of concurrently running simulation threads is to ensure the correct timing order of all simulation events (also called causality). The general parallelization of a *discrete-event simulation* (DES) is known as *parallel discrete-event simulation* (PDES) [3], which refers to the execution of a single DES on a parallel machine. There are mainly two algorithm classes proposed to solve the synchronization problem: *Conservative algorithms*, which guarantee by design that no event can occur in the past of an already advanced simulation thread and *optimistic algorithms*, which allow simulation threads to advance asynchronously and provide rollback mechanisms to certain simulation states, if an event happened in the past. Our presented parallelization techniques are conservative algorithms. Thus, simulation threads are only allowed to be activated and to execute simulation tasks if causality is ensured. In the following sections, we present different parallelization techniques that enable parallel simulation and always ensure causality.

5.1.2 *Parallelization Techniques*

In this section, we explain how we parallelized our direct-execution simulation approach. Parallel simulation requires a couple of global data structures and variables that are accessible by all simulation threads. The global data structures are:

- T—Min-heap that contains all simulation tasks ordered by their simulation times
- R—Set of currently running simulation threads
- W—Set of waiting simulation threads

The min-heap provides the functions PUSH to add a new simulation task onto it, GETFIRST to return and POPFIRST to remove the simulation task with the globally smallest simulation time (all functions ensure the ordering of the min-heap). The initialization of the simulation and the main loop of a simulation thread are shown in Algorithm 5.1. At the beginning of the simulation, the data structures are initialized and the first simulation task is pushed onto the min-heap. The variable *global_time* represents the globally smallest simulation time among all simulating simulation threads and is initialized with 0. The *terminated* flag indicates whether the entire simulation is finished and all simulation threads may return. The number of parallel simulation threads to be started is specified by the variable n. In line 6, n simulation threads are created by calling CREATETHREAD (this function creates an X10 worker thread). Each simulation thread then executes a main loop by calling RUNTHREAD. The simulation thread d is added to the set of currently running simulation threads R and executes simulation tasks by calling EXECUTETASK as long as the *terminated* flag is not set. An execution of a simulation task can be only started if the condition in line 15 is satisfied. The simulation time of the simulation task returned by GETFIRST needs to be equal to *global_time*, otherwise causality cannot be guaranteed. If the condition is not true, this simulation thread is sent to sleep by calling SENDTHREADTOSLEEP, because it cannot be used for parallel execution currently (it is removed from the set R and added to the set W). If the condition is true, the simulation task is removed from the min-heap by POPFIRST and stored locally at the simulation thread (accessible via *task*(d)). Functional simulation is activated by calling STARTFUNCTIONALSIMULATION and the actual application activity is started by RUNACTIVITY (internally, a context switch is applied). During the execution of the activity, new activities may

Algorithm 5.1 Initialization and simulation thread main loop.

```
1  PUSH(T, task_0)                                    ▷ T – Min-heap of simulation tasks
2  R ← ∅                                              ▷ R – Set of running simulation threads
3  W ← ∅                                              ▷ W – Set of waiting simulation threads
4  global_time ← 0
5  terminated ← false
6  for all i ∈ {0, ..., n − 1} do
7      d_i ← CREATETHREAD
8      RUNTHREAD(d_i)

9  procedure RUNTHREAD(d)
10     R ← R ∪ d
11     while ¬terminated do
12         EXECUTETASK(d)
13     R ← R \ d

14 procedure EXECUTETASK(d)
15     if |T| > 0 ∧ time(GETFIRST(T)) = global_time then
16         task(d) ← POPFIRST(T)
17         STARTFUNCTIONALSIMULATION(d)
18         RUNACTIVITY(activity(task(d)))
19         STOPFUNCTIONALSIMULATION(d)
20         SYNCHRONIZE(d)
21         if |R| = 1 then
22             if |T| = 0 then
23                 terminated = true
24                 for all d' ∈ W do
25                     W ← W \ d',  R ← R ∪ d'
26                     WAKEUPTHREAD(d')
27             else
28                 global_time ← time(GETFIRST(T))
29     else
30         R ← R \ d,  W ← W ∪ d
31         SENDTHREADTOSLEEP(d)
```

be created and thus, new simulation tasks are pushed into the min-heap. After the activity finishes, the functional simulation is stopped using STOPFUNCTIONALSIMULATION, which internally applies time warping to advance the simulation time of the simulation task. Synchronization with the other simulation tasks is required by calling SYNCHRONIZE (see Algorithm 5.2), because the termination of an activity is a synchronization point. The procedures for starting and stopping the functional simulation are shown in Algorithm 5.4. After synchronization, the termination condition is checked in line 21 and 22. If only one simulation thread is currently running and no further simulation task is available in the min-heap, the entire simulation is finished and the *terminated* flag is set to *true*. In this case, if other simulation threads are still waiting, each is awakened by calling WAKEUPTHREAD (and removed from the set W and added to set R). They will check the *terminated* flag and return. If further simulation tasks exist in the min-heap, the global simulation time is advanced to the

Algorithm 5.2 Synchronization and simulation thread activation.

```
1  procedure SYNCHRONIZE(d)
2     global_time ← GETGLOBALMINTIME
3     WAKEUPTHREADS
4     if time(task(d)) > global_time then
5        PUSH(T, task(d))
6        EXECUTETASK(d)

7  procedure WAKEUPTHREADS
8     if |T| > 0 ∧ time(GETFIRST(T)) = global_time then
9        for all d′ ∈ W do
10          W ← W \ d′,  R ← R ∪ d′
11          WAKEUPTHREAD(d′)
```

Algorithm 5.3 Determination of the minimum simulation time.

```
1  function GETGLOBALMINTIME
2     min_time ← min_{∀d′∈R} {time(task(d′))}
3     if |T| > 0 then
4        min_time ← min(time(GETFIRST(T)), min_time)
5     return min_time
```

time of the first simulation task from the min-heap in line 28 and it will be executed next by the simulation thread.

During the simulation of an activity, it might enter a synchronization point (e.g., creating a new activity), which requires synchronization with other simulation tasks in order to guarantee causality. After functional simulation has been stopped, the simulation time of the simulation task is updated (using time warping) and SYNCHRONIZE in Algorithm 5.2 is called. In SYNCHRONIZE, the smallest simulation time among all simulation tasks of currently running simulation threads (set R) and the first simulation task on the min-heap is determined by calling GETGLOBALMINTIME (see Algorithm 5.3). The global simulation time *global_time* is updated accordingly. If the smallest simulation time is determined by the first simulation task on the min-heap, another simulation thread may be started. Thus, WAKEUPTHREADS is called, which checks the condition for a simulation task to be executed in line 8 and if true, awakes all waiting simulation threads (set W) by calling WAKEUPTHREAD for each. The access to the min-heap is safe against race conditions. Hence, only one simulation thread may be successful in popping the simulation task from it. If multiple simulation tasks have the smallest simulation time, multiple simulation threads may successfully pop one of them. In line 4, it is checked whether the current simulation thread may continue the execution of its current simulation task. This is only allowed if the simulation time of the simulation task is equal to *global_time*. Otherwise, this simulation task is pushed onto the min-heap by calling PUSH and another simulation task may be executed by calling EXECUTETASK from Algorithm 5.1.

Starting and stopping functional simulation of an activity running on a simulated target core is shown in Algorithm 5.4. As explained in Sect. 3.3.1, our time warping

Algorithm 5.4 Starting and stopping functional simulation.

1 **procedure** StartFunctionalSimulation(d)
2 ResetPC($pc(d)$)

3 **procedure** StopFunctionalSimulation(d)
4 $(c_0, c_1, c_2) \leftarrow$ ReadPC($pc(d)$)
5 $(p_0, p_1, p_2) \leftarrow$ GetWeights($core(task(d))$)
6 $f \leftarrow$ GetFrequency($core(task(d))$)
7 $\Delta t \leftarrow$ TimeWarping($(c_0, c_1, c_2), (p_0, p_1, p_2), f$)
8 $time(task(d)) \leftarrow time(task(d)) + \Delta t$

mechanism is based on the performance counter values c_0, c_1, and c_2. Once functional simulation is started, the performance counter $pc(d)$ of the current simulation thread d has to be reset using ResetPC in line 2. During activity execution, they count the corresponding events on the host machine. If functional simulation is stopped, the number of performance counter events is read using ReadPC and stored in the tuple (c_0, c_1, c_2) in line 4. In order to apply the time warping estimation, the performance counter weights as well as the clock frequency from the simulated target core have to be determined. This is done by calling GetWeights and GetFrequency on the currently simulated target core $core(task(d))$ of simulation task $task(d)$. The performance counter weights are stored in the tuple (p_0, p_1, p_2) and the core clock frequency in f. Finally, TimeWarping is called by providing all required arguments and the resulting estimated execution time Δt for the last code section is calculated. The simulation time of the current simulation task is advanced by Δt in line 8.

5.1.2.1 Signaling

Our first parallelization approach is called *signaling*. Here, a simulation thread receives a notification (signal) once it has passed the simulation time of the first simulation task on the min-heap. In this case, another simulation thread may start executing the first simulation task from the min-heap, because the causality condition is satisfied. In order to receive a notification after a certain amount of simulation time has expired, we exploit the *sampling* and *signaling* mechanism of performance counters. This mechanism allows programming a performance counter to send a UNIX signal to a simulation thread after a specific amount of events have been counted (e.g., after a certain amount of retired instructions have been executed).

The signaling technique is shown in Algorithm 5.5. The first part of the algorithm is supposed to extend StartFunctionalSimulation from Algorithm 5.4. It is inserted between line 1 and 2 of Algorithm 5.4. Thus, the signaling concept is applied, when a simulation thread starts function simulation of a simulation task by calling StartFunctionalSimulation. Only if further simulation tasks exist in the min-heap and waiting simulation threads are available (see line 2 of Algorithm 5.5), signaling is applied. The time difference Δt between the time of the first simulation

Algorithm 5.5 Signaling technique.

```
1  if using signaling then                     ▷ Is inserted between line 1 and 2 of Algorithm 5.4
2    if |T| > 0 ∧ |W| > 0 then
3      Δt ← time(GETFIRST(T)) − global_time
4      (p0, p1, p2) ← GETWEIGHTS(core(task(d)))
5      f ← GETFREQUENCY(core(task(d)))
6      c0 ← REVERSETIMEWARPING(Δt, p0, f)
7      PROGRAMPC(pc(d), c0)

8  procedure PCSIGNALHANDLER(d)
9    STOPFUNCTIONALSIMULATION(d)
10   global_time ← GETGLOBALMINTIME
11   WAKEUPTHREADS
12   STARTFUNCTIONALSIMULATION(d)
```

task of the min-heap and *global_time* is determined, because this is the next event, which could be executed by another simulation thread. Now, the problem is how to determine that the functional simulation of the current activity on a simulated target core has passed that point in simulation time. Here, we apply the inverse operation of time warping called *reverse time warping* in order to determine an amount of performance counter events on the host machine that would result in the simulation time delta Δt if time warping according to Eq. (3.1) is applied. The performance counter is then configured to send a notification after the determined amount of events has been counted. Since reverse time warping cannot provide a unique solution for all performance counter events c_0, c_1, and c_2, we apply an over-approximation by assuming c_1 and c_2 being 0. We can determine a unique solution for the number of retired instructions c_0 according to the following equation:

$$c_0 = \Delta t \cdot f \cdot \frac{1}{p_0} \tag{5.1}$$

This assumes that no memory accesses are executed and only retired instructions account for the simulation time delta Δt. It is an over-approximation since the weights for retired instructions are typically smaller than for the memory accesses. Thus, a notification will safely occur after the simulation time of the next event. In Algorithm 5.5, the lines 4 to 6 determine the performance counter weights and the clock frequency of the current target core *core*(*task*(*d*)) and call REVERSETIMEWARPING, which applies Eq. (5.1) to determine the number of retired instructions c_0 that have to be executed on the host machine after a notification is sent. The performance counter *pc*(*d*) is programmed accordingly using PROGRAMPC. Once the configured amount of retired instructions have been executed by the activity, the performance counter will send a signal to the simulation thread, which is handled by the signal handler PCSIGNALHANDLER shown in Algorithm 5.5. This handler temporarily stops functional simulation to update the current simulation time of the simulation task. The global simulation time *global_time* is updated in line 10 and WAKEUPTHREADS is

Algorithm 5.6 Luck-shot technique.

```
if using luck-shot then                                  ▷ Replaces line 2 of Algorithm 5.3
    for all d' ∈ R do
        (c_0, c_1, c_2) ← READPC(pc(d'))
        (p_0, p_1, p_2) ← GETWEIGHTS(core(task(d')))
        f ← GETFREQUENCY(core(task(d')))
        Δt ← TIMEWARPING((c_0, c_1, c_2), (p_0, p_1, p_2), f)
        min_time ← min(time(task(d')) + Δt, min_time)
```

called in line 11. If not other simulation thread is currently running, the global simulation time is advanced to the simulation time of the next event in the min-heap and another simulation thread will concurrently execute the corresponding simulation task.

An example execution trace of the signaling parallelization technique is depicted in Fig. 5.2. Simulated are three target CPUs as shown in the upper part. Three simulation threads are available on the host machine and shown in the lower part. Assuming each target CPU entered a synchronization point at the times t_1, t_2, and t_3 on the timeline (depicted as blue dot). Then, only one simulation thread is allowed to continue. Simulation thread 2 is selected and starts functional simulation of CPU 2. The performance counter of simulation thread 2 is programmed according to the time difference $\Delta t = t_2 - t_1$. Then, functional simulation of CPU 2 starts. Once the simulation thread is signaled, it starts simulation thread 1, which then simulates CPU 1 in parallel to simulation thread 2. Now, both simulation threads program their performance counters according to the time difference $\Delta t = t_3 - t_2$ and start functional simulation. Only if both simulation threads got signaled, simulation thread 0 can be started for simulating CPU 0 in parallel. If all n available simulation threads are running in parallel, no signaling is applied since no further simulation thread is available for parallel execution. In this case, all simulation threads have to simulate until they enter their next synchronization point. If further simulation tasks are available, they are executed, otherwise all simulation threads have to wait until all of them entered a synchronization point before functional simulation can be safely continued. This might drastically reduce the number of simultaneously running simulation threads.

5.1.2.2 Luck-Shot

The disadvantage of the signaling technique described before can be improved by a technique to determine a snapshot of the current simulation time during functional simulation. Such snapshots may be acquired by reading the current amount of events counted by a performance counter of a running simulation thread and then applying the time warping mechanism to calculate the current simulation time of this simulation thread. We call this technique *luck-shot* and it is depicted in Algorithm 5.6. It is supposed to be used in the GETGLOBALMINTIME call from Algorithm 5.3 to replace the minimum simulation time calculation of the currently running simulation threads

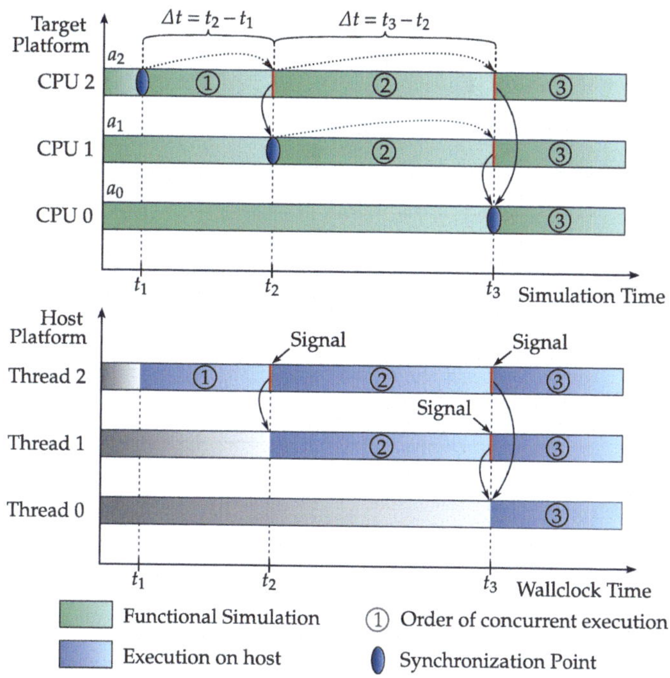

Fig. 5.2 Signaling-based parallelization technique. Three target CPUs are simulated and three simulation threads are available on the host platform [4]

(set R). If a simulation thread enters a synchronization point, it calls SYNCHRONIZE from Algorithm 5.2. Then, the global simulation time is determined by evaluating the simulation times of the currently running simulation threads. Without using the luck-shot technique, the simulation times from the last events of each simulation thread are evaluated. When using the luck-shot technique, for each currently running simulation thread, a simulation time snapshot is determined in lines 3 to 7 of Algorithm 5.6 and used for minimum simulation time calculation. If this evaluation results in a global simulation time that is bigger than the simulation time of the current simulation thread (see line 4 in Algorithm 5.2), functional simulation is continued immediately. Otherwise, the simulation thread is going to be blocked if not further simulation task is available. However, there is no guarantee that the other simulation threads have passed the simulation time of the current simulation thread.

In Fig. 5.3, an example execution trace of the luck-shot parallelization technique is depicted. Two target CPUs are simulated as shown in the upper part and two simulation threads are available on the host machine as shown in the lower part. Initially, both simulation threads are running and CPU 1 enters a synchronization point (depicted as blue dot) at time t_1. Now, a simulation time snapshot of simulation thread 0 is determined that is based on its current performance counter values. The evaluation resulted in a bigger simulation time than from simulation thread 1. Thus,

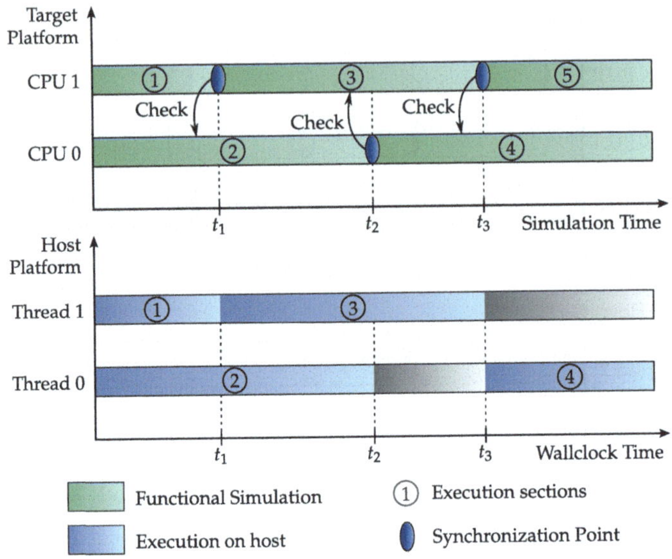

Fig. 5.3 Luck-shot parallelization technique. Two target CPUs are simulated and two simulation threads are available on the host platform [4]

functional simulation of CPU 1 is continued immediately. Next, CPU 0 enters a synchronization point at time t_2 and determines a snapshot of the current simulation time of simulation thread 1. The evaluation resulted in a smaller simulation time than from simulation thread 0. Thus, simulation thread 0 is blocked. Once CPU 1 enters a synchronization point at time t_3, simulation thread 0 is awakened, because it has the smaller simulation time and simulation thread 1 is blocked.

5.1.2.3 Heartbeat

Our last parallelization technique is called *heartbeat*, which is also based on simulation time snapshots, similar to the luck-shot technique. Here, a dedicated X10 worker thread (the heartbeat thread) is spawned at the beginning of the simulation and periodically triggered to check the simulation time snapshots of the currently running simulation threads and to decide whether any waiting simulation thread may be activated for parallel simulation. The technique is presented in Algorithm 5.7. The first part of the algorithm is appended to the initialization of the simulation after line 8 of Algorithm 5.1. Here, an X10 worker thread is created in line 2 using CREATETHREAD and starts its execution by calling RUNHEARTBEATTHREAD. This X10 worker thread runs concurrently to the simulation threads until the *terminated* flag is set to *true*. The heartbeat thread sleeps for a given period of time p in line 8. If the heartbeat thread returned from sleep, it determines the smallest global simulation time of by calling GETGLOBALMINTIME. In order to check the current simulation

Algorithm 5.7 Heartbeat technique.

```
if using heartbeat then                              ▷ Is appended after line 8 of Algorithm 5.1
    d ← CREATETHREAD
    RUNHEARTBEATTHREAD(d, p)

procedure RUNHEARTBEATTHREAD(d, p)
    while ¬terminated do
        global_time ← GETGLOBALMINTIME
        WAKEUPTHREADS
        SENDTHREADTOSLEEP(d, p)
```

time snapshots of the running simulation threads, the luck-shot technique is required. If the global simulation time is equal to the time of the next event of the min-heap, a waiting thread may be started for parallel simulation by calling WAKEUPTHREADS. Afterwards, the heartbeat thread again sleeps for the period of time p.

An example execution trace of the heartbeat parallelization technique is depicted in Fig. 5.4. Two target CPUs are simulated as shown in the upper part and two simulation threads are available on the host machine as shown in the lower part. Assuming each target CPU entered a synchronization point at the times t_1 and t_2 on the timeline (depicted as blue dot). Since t_1 is smaller than t_2, simulation thread 1 continues functional simulation. During the execution of simulation thread 1, the heartbeat thread is activated at time b_1. Since the simulation time of simulation thread 1 is not yet bigger or equal than t_2, simulation thread 0 is not awakened. At the next activation of the heartbeat thread at time b_2, the simulation time of simulation thread 1 is bigger than t_1 and simulation thread 0 is started. Since CPU 1 enters a synchronization point at time t_3, simulation thread 1 is stopped and also not activated at the next heartbeat activation at time b_3.

The selection of an appropriate heartbeat period p is essential for the performance of this parallelization technique since simulation threads are only started at these periodic times. If the period p is chosen too small (e.g., smaller than 100 µs), the overhead of activating and deactivating the heartbeat thread may increase significantly. If the period p is chosen too big (e.g., bigger than 100 ms), waiting simulation threads might be unnecessarily delayed.

5.1.2.4 Adaptive Heartbeat

As described before, the heartbeat period p has to be selected carefully. For the presented heartbeat techniques, the period p is not changed at run time. In this section, we present an *adaptive* heartbeat technique, where the heartbeat period is changed at run time according to the dynamically changing parallel behavior of the simulated applications to maximize the speedup gained through parallel simulation. In Fig. 5.5, we evaluated different heartbeat periods p ranging from 100 µs to 250 ms regarding simulation latency for a selected simulation scenario. A distributed FFT

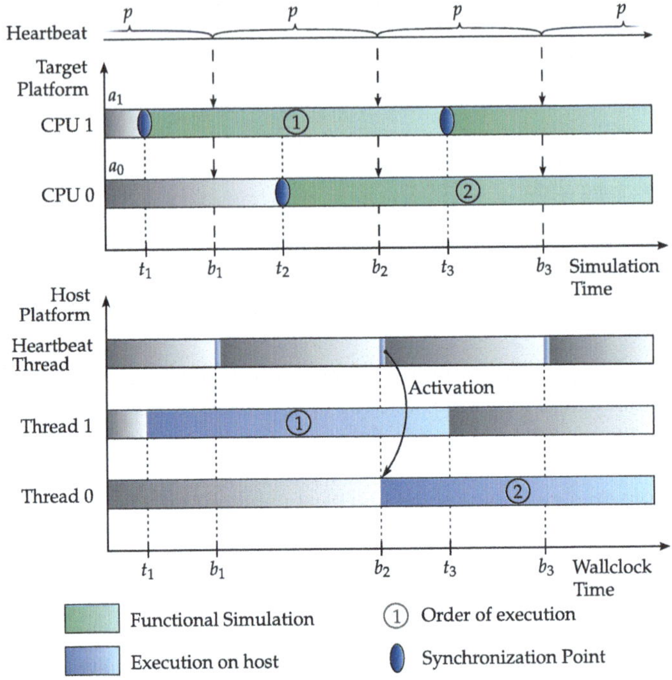

Fig. 5.4 Heartbeat parallelization technique. Two target CPUs are simulated and two simulation threads are available on the host platform [4]

application is simulated on a 12-tile architecture. The number of host cores available for parallel simulation varies from 1 to 6. In case of one host core, it can be seen that the simulation latency increases continuously with a decreasing period. For more host cores, the simulation latency decreases with a decreasing period and increases again if the period has been chosen too small. However, the sensitivity regarding very small periods is getting lower with an increasing number of available host cores since the overhead of the heartbeat thread affects only a single host core.

According to this evaluation, we developed a simple heuristic of a dynamic period adaption in order to maximize the speedup through parallel simulation. At each heartbeat activation, the average CPU usage of the running simulation threads is determined for the last period and compared to the CPU usage of the heartbeat thread for the last period. If the CPU usage of the heartbeat thread is greater than the average CPU usage of the running simulation threads and the magnitude of the difference is greater than a certain threshold, the period is increased by a certain increment in order to reduce the CPU usage of the heartbeat thread. If the CPU usage of the heartbeat thread is smaller, the period is decreased to increase it again. The operating points of the adaptive heartbeat technique for the given simulation scenario are marked by red circles in Fig. 5.5.

Fig. 5.5 Evaluation of the heartbeat period $p$for a simulation of an FFT application scenario in dependence on the number of cores available on the host for parallel simulation [1]

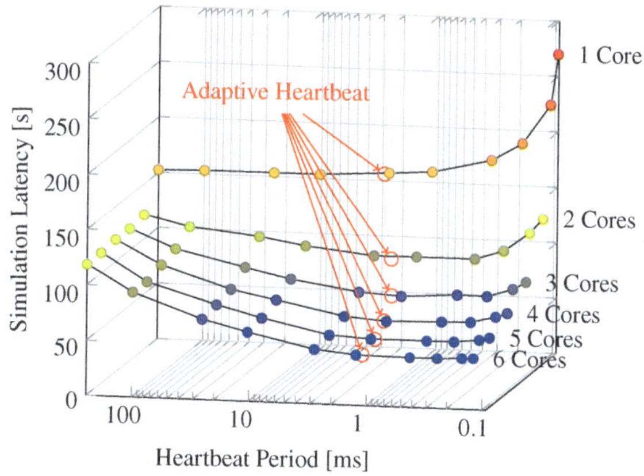

5.1.3 *Experimental Results*

In the following experiments, we evaluate the achievable speedup of our proposed four parallel simulation approaches on a multi-core host machine for different parallel application workloads running on a tiled multi-core architecture. In order to exploit parallel simulation, the simulated application and architecture must provide a certain degree of parallelism, otherwise the simulation cannot be accelerated by multi-core host machines. One approach would be to simulate completely independent instances of the same or different applications, each on a different tile. However, this would represent an ideal world with no communication or synchronization between the applications, which is not realistic. Thus, we used an existing X10 benchmark suite, which provides real-world parallel applications from different domains such as signal processing (FFT), graph analysis, k-Means optimization, and streaming applications. These are perfect candidates for evaluating our parallel simulation approaches on a multi-tile MPSoC architecture, because they make extensively use of distributed memory parallelism including communication and synchronization. Simulating the communication overhead over the NoC is expensive in the sense that a lot of synchronization points are generated due to locking, creating new activities on other places, waiting for activities, and sending back replies.

We evaluated our different parallelization techniques by simulating the presented applications on a simulated homogeneous 16-tile architecture (4×4) with each tile having 4 RISC cores and measuring the resulting speedup of the parallel simulation compared to sequential simulation. The host machine provides two Intel Xeon processors each having 6 physical cores. Although, the cores provide $2\times$ *simultaneous multi-threading* (SMT), our simulations could not benefit from the hardware-based multi-threading. We pinned the simulation threads to the physical cores and neglected the additional SMP cores. Thus, we could use a maximum of 12 cores in total for the parallel simulation. The results of this evaluation is shown in Fig. 5.6.

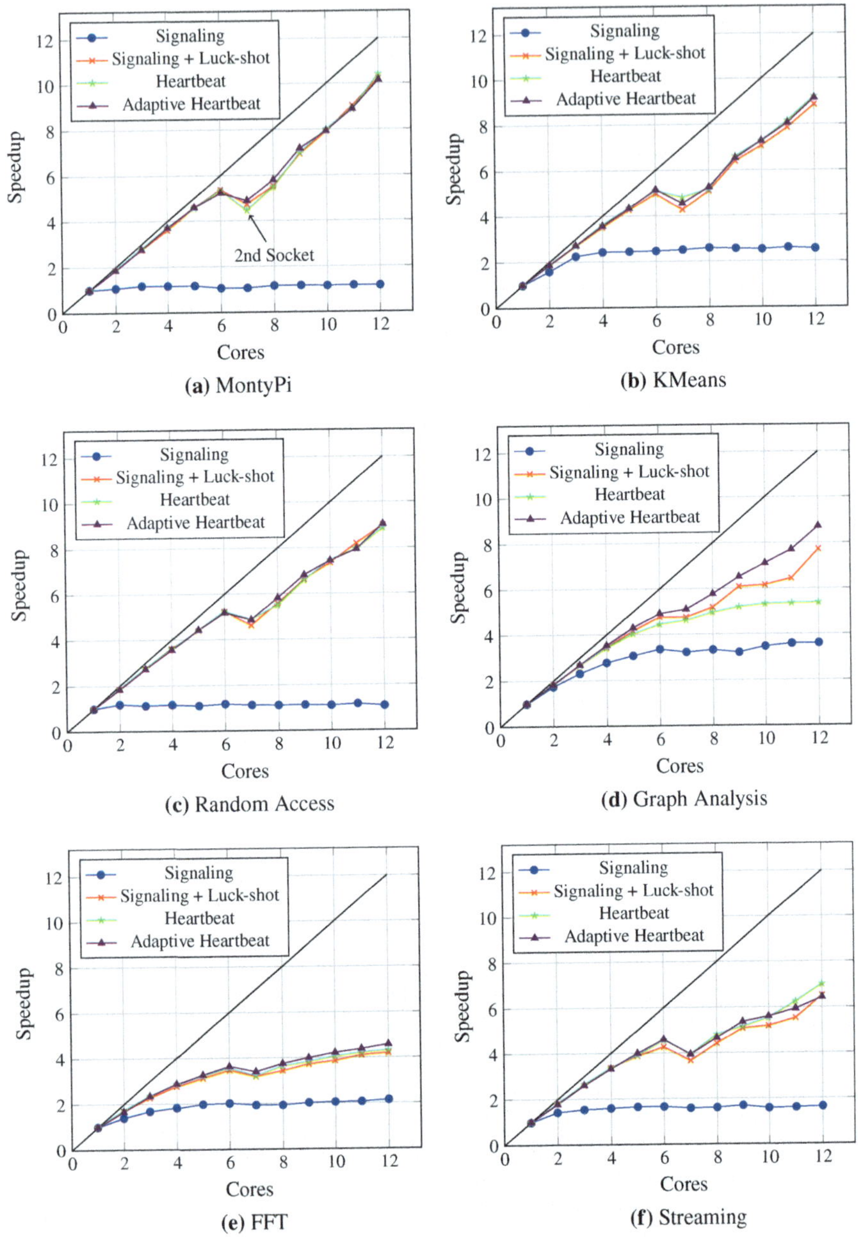

(a) MontyPi (b) KMeans (c) Random Access (d) Graph Analysis (e) FFT (f) Streaming

Fig. 5.6 Evaluation of an X10 benchmark suite of several distributed-parallel applications [1]

We achieved an average speedup of 8.2 compared to sequential simulation. The applications scaled well with an increased number of used host cores for the parallel simulation. For all applications, a slight drop of the speedup can be observed for more than 6 host cores. The reason is the usage of the second processor socket of the Intel Xeon host machine, which requires inter-socket communication. As one can see in the charts, the signaling parallelization technique without luck-shot does not scale very well. This is mainly because of the already discussed drawback in Sect. 5.1.1. We figured out that the used locks in the applications result in synchronization points, which follow each other closely in simulation time (one at the entry of a lock and one at the exit). These small functional simulation sections between synchronization points in combination with huge functional simulation sections of other simulation threads slow down the signaling parallelization technique drastically. The luck-shot parallelization technique improves signaling notably as can be seen in the charts. It provides almost the same speedup as the heartbeat technique. The static heartbeat period p was set to $p = 5$ ms, which results in almost linear scaling for almost every application. The FFT calculation does not scale well for any of the parallelization techniques, because it heavily makes use of expensive all-to-all communication primitives, which results in a lot of synchronization points. The adaptive heartbeat techniques still improves the results of the static-period heartbeat technique, especially in the case of the graph analysis application. Here, even the signaling technique with luck-shot performs better than the static-period heartbeat technique. Employing 12 host cores of our simulation host machine results in a total simulation performance of up to 86,000 *millions instructions per second* (MIPS).

5.2 Architecture Evaluation Case Studies

In the last section, we introduced parallelization concepts for the direct-execution simulation approach of InvadeSIM to exploit the performance of multi-core host processors for complex simulations. We will use this potential in this section for architecture evaluation. As introduced earlier, processor architectures are becoming not only more and more parallel but also increasingly heterogeneous for energy efficiency reasons. For reasons of scalability, tiled MPSoC architectures have been proposed for implementing many performance and power-critical domain-specific systems. This trend toward many-core system designs implementing hundreds to thousands of cores as well as hardware accelerators on a single chip leads to many different challenges such as overheating, reliability and security issues as well as resource contentions. Here, resource-aware programming concepts such as through *invasive computing* [5, 6] have been proposed that exploit self-adaptiveness and self-organization in resource allocation and distribution of workload to tackle these issues. However, another obstacle of such architectures is the aforementioned heterogeneity. It poses a big problem on how to evaluate architectural design choices early such as number of tiles, internal tile structure and selection of cores within each tile in order to provide the best implementation for a set of given applications. Thus, in this section, we present different case studies on evaluating sets of resource-aware applications on heterogeneous MPSoCs to compare different invasive computing architecture designs. For an efficient evaluation of such important architectural decisions, InvadeSIM is used since it provides fast and flexible simulation of heterogeneous architectures. For selected parallel application scenarios, we show how architectural parameters may be varied very fast, which in turn enables the exploration of different designs for cost, performance, and other design objectives.

5.2.1 Architecture Description

Our considered MPSoC architectures are tiled and heterogeneous as already introduced in Sect. 2.1. Figure 5.7 displays a typical example of such an MPSoC architecture as considered in the following. Shown is a generic tiled architecture, which consists of several heterogeneous tiles, including computing tiles with different types of processor cores, memory tiles for global memory accesses, tiles with dedicated accelerator such as TCPAs, and I/O tiles. The tiles are assumed to be interconnected in a grid-like topology by a network-on-chip. Each tile is equipped with NA, which acts as interface to the NoC. For domain-specific customization, we characterize a tiled architecture by the following structural parameters:

- The NoC width X^{noc}, height Y^{noc}, and frequency f^{noc}
- A set of $n = X^{noc} \cdot Y^{noc}$ tiles $\mathbf{T} = \{T_1, \ldots, T_n\}$
- Each tile $T_i \in \mathbf{T}$ is associated with the following parameters to specify its internal structure:
  - A set of s_i cores $\mathbf{C}_i = \{C_{i,1}, \ldots, C_{i,s_i}\}$
  - A set of p_i TCPAs $\mathbf{A}_i = \{A_{i,1}, \ldots, A_{i,p_i}\}$ (for p_i typically applies $0 \leq p_i \leq 1$)
- Each core $C_{i,j} \in \mathbf{C}_i$ is associated with the following parameters:
  - A core type $r_{i,j} \in \{\text{RISC}, i\text{-Core}\}$
  - A core frequency $f_{i,j}^{core}$
- Each TCPA $A_{i,k} \in \mathbf{A}_i$ is associated with the following parameters:
  - A TCPA width $X_{i,k}^{tcpa}$ and height $Y_{i,k}^{tcpa}$
  - A TCPA frequency $f_{i,k}^{tcpa}$
- Furthermore, each resource type is associated with specific costs (e.g., FPGA hardware resources)
  - Costs for a tile c^{tile} include the costs for the TLM, the NA, and a NoC router
  - Costs for a RISC core c^{risc} and the i-Core c^{icore}
  - Specific costs $c^{tcpa_{pe}}$ for a single TCPA PE

For example, the tiled structure of the architecture from Fig. 5.7 may be described as follows: $X^{noc} = 3$, $Y^{noc} = 3$, $n = 9$, and $\mathbf{T} = \{T_1, T_2, T_3, T_4, T_5, T_6, T_7, T_8, T_9\}$.

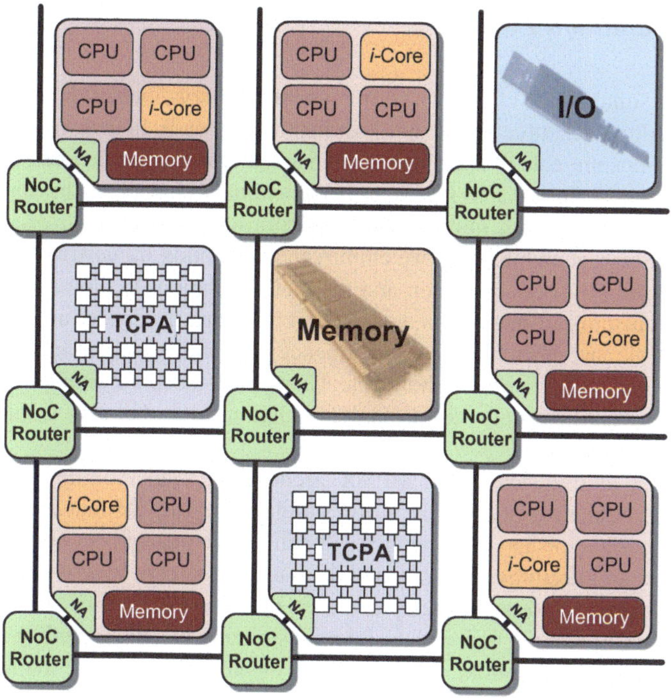

Fig. 5.7 A generic invasive tiled architecture with different processor types, accelerator tiles such as TCPAs, and memory tiles connected by a NoC [7]

For instance, the upper left tile T_1 can be described as follows: $s_1 = 4$, $p_1 = 0$, $\mathbf{C}_1 = \{C_{1,1}, C_{1,2}, C_{1,3}, C_{1,4}\}$, and $\mathbf{A}_1 = \emptyset$. The type of the cores are: $r_{1,1} = \text{RISC}$, $r_{1,2} = \text{RISC}$, $r_{1,3} = \text{RISC}$, and $r_{1,4} = i\text{-Core}$. Similarly, the upper TCPA tile T_4 can be characterized by $s_4 = 1$, $p_4 = 1$, $\mathbf{C}_4 = \{C_{4,1}\}$, and $\mathbf{A}_4 = \{A_{4,1}\}$. Each TCPA tile always contains a RISC core for controlling the TCPA execution, which is not shown in Fig. 5.7. The core of the TCPA tile is specified by: $r_{4,1} = \text{RISC}$ and the TCPA by: $X_{4,1}^{tcpa} = 6$ and $Y_{4,1}^{tcpa} = 5$. We developed an architecture description XML file that allows specifying the introduced architecture structure and parameters. It allows evaluating different architecture configurations with InvadeSIM quickly. In Sect. 5.2.3, we will provide respective cost and frequency values for the components of our considered architectures and subsequently present evaluation results for cost/performance tradeoffs based on our simulations with InvadeSIM.

5.2.2 Evaluation Flow

We want to simulate the effect of varying resource reservations of an MPSoC platform for multiple resource-aware applications running on it. In invasive computing, the applications are able to specify their dynamically changing demand of parallel computing power. In order to simulate the functional behavior of the applications as well as to receive back important timing information about their execution on the simulated target architecture, we employ the simulation framework InvadeSIM, which is based on direct-execution and parallel simulation. It allows creating and modifying customized heterogeneous tiled architectures easily by an architecture description file and evaluating the performance of architecture variants apart from functional correctness in a very short time. Using cycle-accurate simulators for timing analysis would not only lead to long simulation times, but also would require complex hardware models.

In Fig. 5.8, an overview of our MPSoC architecture evaluation flow is depicted. The central component of this flow is the simulator InvadeSIM that allows for fast functional verification and performance estimation. We manually configured different hardware architectures and evaluated them against a set of invasive applications using InvadeSIM. Important design parameters that have been varied and are of high interest are the number of tiles and the internal structure of a tile including the selection of cores and accelerators. The objectives of our evaluation are the total costs of an architecture variant as well as different performance metrics to compare the architectures. The total costs of an architecture are calculated by summing up the costs of all contained resources in that architecture such as tiles (including routers), cores (RISC and *i*-Core), and TCPAs. Our considered invasive applications compete against each other for resources, employ different degrees of parallelism, and exploit different hardware accelerators such as an *i*-Core or a TCPA for speeding up their computations.

5.2.3 Experimental Results

In our experiments, we evaluate different configurations of homogeneous and heterogeneous architectures against costs and performance using InvadeSIM. The simulation runs have been executed on an Intel Core i7 quad-core host CPU with eight virtual cores at 2.93 GHz. In Table 5.1, the costs in terms of required area (determined by the area overhead for a synthesized FPGA implementation) and the static frequencies for the different types of resources of a heterogeneous architecture are depicted. We have chosen the distributed parallel k-Means application from Sect. 5.1.3 and modified it to use resource-aware programming constructs and to exploit different accelerators if available. At the beginning of the application, it invades a configurable number of cores (which may be on different tiles), spreads its workload onto them to process it in parallel, and finally retreats from the processing resources if the computations are finished.

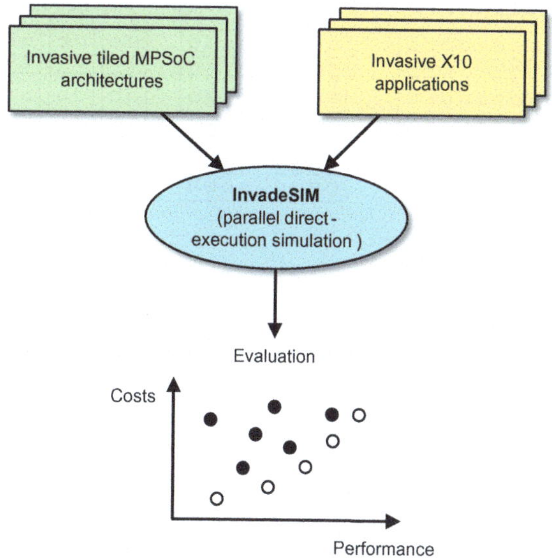

Fig. 5.8 Architecture evaluation flow (adapted from [2])

Table 5.1 Costs in terms of required area and clock frequencies of the different MPSoC resource types used in the experiments

Resource type	Costs (area units)	Clock frequency (MHz)
RISC core	55	500
i-Core	300	500
TCPA PE	13	200
Tile	34	–

5.2.3.1 Homogeneous Architectures

Our first evaluation aspect is the tile layout of the target MPSoC. Here, we considered homogeneous architectures comprised in total of 64 RISC cores arranged in different tile layouts. Figure 5.9 depicts the different variants of the layouts we evaluated in our experiments. First, we simulated only a single k-Means application on four architecture variants according to these tile layouts and measured its execution time (latency). We increased the number of invaded processing resources for the application continually. The results are shown in Fig. 5.10. Here, one can see that the latency of the application on all architecture variants is first reduced by claiming more processor cores for parallel execution, but then is increased again if a large amount of processing resources is claimed. This saturation is caused by the parallel accesses to the shared memory. The results show that the tile layout 1×64 results in a higher latency than the other tile layouts when more cores are claimed, because the bandwidth limitation on a shared memory system reduces the execution time of all activities on the tile. Among the four different considered architecture configura-

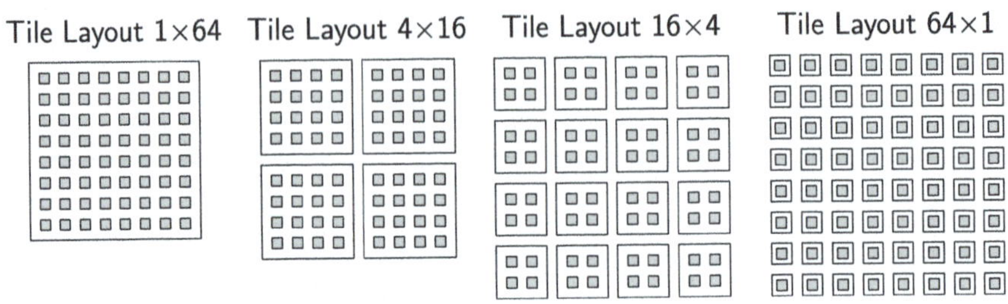

Fig. 5.9 Selected tile layouts of a grid-based homogeneous multi-tile architecture. Each tile consists of one or more equal RISC cores, e.g., the architecture on the left consists of only one tile that contains 64 RISC cores, while the architecture on the right consists of 64 tiles each containing one RISC core [2]

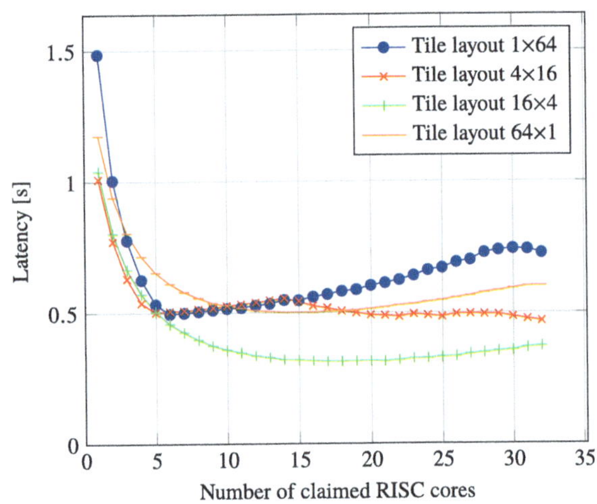

Fig. 5.10 Simulation of a k-Means optimization on a homogeneous architecture with different tile layouts. The computations are spread onto the number of claimed RISC cores and processed in parallel [2]

tions, the best layout for this application is 16 × 4. It represents the best mixture of shared and distributed memory system. Here, the bandwidth limit on one tile only affects a few activities. However, it also runs in a saturation due to the communication overhead between the tiles. The other two tile layouts also represent trade-offs between communication overhead and shared-memory access latency, but result in higher application latencies than the 16 × 4 layout. The 64 × 1 layout results in a high NoC utilization, because each tile holds only a single processor core. The tiles of the 4 × 16 include 16 processor cores, which result in a higher shared-memory access latency than if only 4 processor cores accessing the shared memory.

In our second experiment, not only one, but a given number of instances of the k-Means application is executed on the considered architecture alternatives. For each application instance, the number of claimed RISC cores is fixed to 4. All application instances are started simultaneously and each application instance immediately claimes its required resources and starts parallel execution. We simulated this application scenario on the architecture variants from Fig. 5.9 and determined the total latency to finish all application instances. The results for the different tile layouts are shown in Fig. 5.11a. One can see that tile layout 1 × 64 results in the highest total latencies, because of the shared-memory bandwidth limitation of only one tile. The other layouts result in smaller total latencies, because multiple tiles are used for executing the parallel application instances. They benefit a lot from multiple tiles, because no communication is required between the application instances and the communication overhead within a single application instance invading only 4 RISC cores is very small. Thus, executing each application instance on a tile with a smaller amount of cores is more beneficial than executing more application instances on tiles with a bigger amount of cores. Again, tile layout 16 × 4 results in the best total latencies for this application scenario.

Now, we consider the same application scenario, but evaluate costs/performance trade-offs of the architecture variants for a fixed number of 16 application instances. The costs of an architecture variant are calculated by the sum of the costs of all contained resources according to the values presented in Table 5.1 and the number of resources given by the corresponding tile layout (see Fig. 5.9). The performance of an architecture variant is defined by the number of started application instances (16 in this case) normalized to the total latency to finish all application instances. The calculated results are depicted in Fig. 5.11b. Since each tile requires has fixed hardware overhead (NoC router, NA, local memory, etc.), the total costs of the architecture variants are increasing with an increasing amount of tiles, even though the number of processing resources remains equal. Thus, 64 × 1 layout requires a maximum amount of costs for a homogeneous architecture, because one tile is used per processing resource. The 1 × 64 layout requires the smallest total costs, but also provides the smallest performance. The 16 × 4 layout provides the best performance and moderate costs. The Pareto-optimal architecture layouts are 1 × 64, 4 × 16, and 16 × 4. The 61 × 1 layout is dominated by 16 × 4 layout.

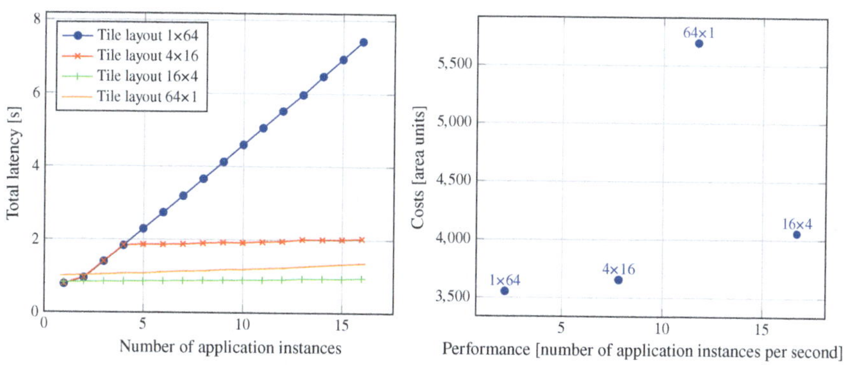

(a) Evaluation of the total latency **(b)** Evaluation of costs/performance trade-offs

Fig. 5.11 Simulation of multiple instances of the k-Means application each claiming four RISC cores for parallel execution on homogeneous architectures with different tile layouts [2]

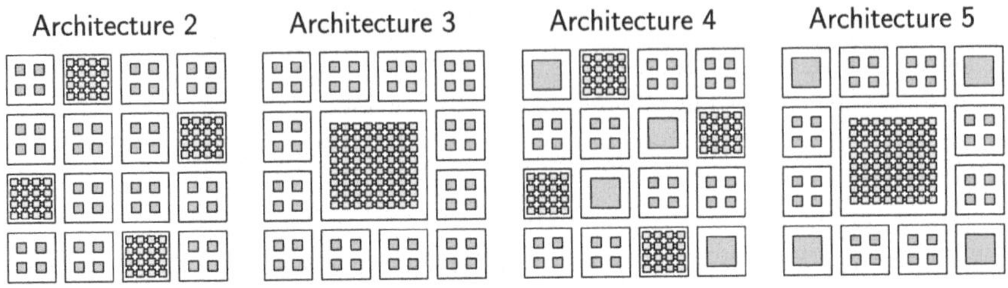

Fig. 5.12 Selected heterogeneous architectures. Each tile may consist either of four RISC cores, one i-Core, a 4×4 TCPA, or an 8×8 TCPA [2]

5.2.3.2 Heterogeneous Architectures

In our second series of experiments, we evaluated the costs and performance of heterogeneous architectures. Here, we studied five different configurations of tiled architectures with a 16×4 layout as shown in Fig. 5.12. They may contain tiles with four RISC cores, one i-Core, a 4×4 TCPA, or an 8×8 TCPA. Architecture 1, which is not shown in this figure is represented by the homogeneous 16×4 architecture as depicted in Fig. 5.9. We modified the k-Means application in order to exploit hardware accelerators such as the i-Core or the TCPA. The application consists of three sub-algorithms with the following characteristics: The first part is a task parallel execution and suitable to be executed in parallel by multiple RISC cores, the second part is computationally intensive and suited for data parallel accelerators such as the TCPA, and the last part may benefit from custom instructions as provided by the i-Core. In this experiment, we only consider executions of k-Means application instances on a single tile. In order to prioritize the invasion of resources for each application instance, we compared the execution times for this application on the different tile types. The largest performance gain could be achieved by using a TCPA tile, because a lot of parallelism can be exploited using 16 or even 64 TCPA PEs. An i-Core tile provides the second best performance and a four core RISC tile provides the worst performance. Thus, at the beginning of a k-Means application instance, an invasion is applied with the following prioritization: (1) An 8×8 TCPA tile, (2) a 4×4 TCPA tile, (3) an i-Core tile, and (4) a tile with four RISC cores. If an invasion of a higher prioritized tile type was not successful, an invasion of the tile type with the next smaller priority is applied.

We simulated a given number of instances of this invasive k-Means application on the considered heterogeneous tiled architectures and determined the individual latencies for each application instance to determine the average latency. The results of this simulation scenario are shown in Fig. 5.13a. It can be seen that the homogeneous architecture (architecture 1) provides a constant average latency for all numbers of application instances. It represents the baseline for the other architecture variants. As can be seen in the diagram, the application instances benefit a lot from the heterogeneous architectures since the average latency is smaller in all cases compared to the homogeneous architecture. Architecture 4 provides the smallest average latency for

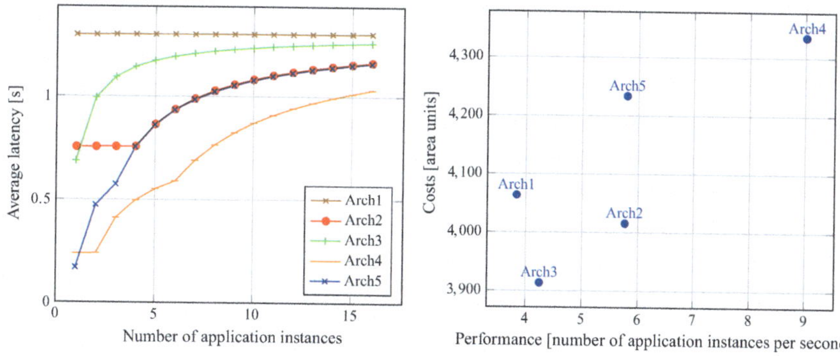

(a) Evaluation of the average latency **(b)** Evaluation of costs/performance trade-offs

Fig. 5.13 Simulation of multiple instances of the invasive k-Means application on heterogeneous architectures with different resource types [2]

the application instances, which is a mixture of small TCPA tiles and i-Core tiles. Those architecture variants with a large TCPA tile (architecture 3 and 5) result in higher latencies than the variants with smaller TCPA tiles.

Again, we consider the same application scenario, but evaluate costs/performance trade-offs of the architecture variants for a fixed number of 16 instances of the invasive k-Means application. Figure 5.13b shows the evaluation results. Costs and performance are determined as explained before. It can be seen that the homogeneous architecture is already dominated by architecture 2 and 3. Both architectures include TCPAs and require less costs and provide better performance than a homogeneous architecture. Furthermore, architecture 2 is more expensive in terms of costs, but also provides a better performance than architecture 3. These results demonstrate that fewer large TCPAs may be used if area costs have to be saved and performance is not of high importance. If performance is more important, multiple smaller TCPAs are preferable. The most expensive, but also the most powerful configurations are architecture 4 and 5. They employ a mixture of i-Cores and TCPAs and require more area costs than a homogeneous architecture. These experiments demonstrate that TCPAs in invasive architectures help to improve the performance, while saving area costs. The usage of i-Cores in invasive architectures further improves the performance, but at the expense of higher area costs.

5.3 Related Work

Different simulation techniques and abstraction levels have been proposed to improve and accelerate simulation. The first class are conventional sequential simulators and emulators. Popular examples in this category are QEMU [8], Simics [9], SimpleScalar [10], SimNow [11], PTLsim [12], and M5 [13]. Some of them are full-system simulators and are also able to model parameterizable multi-core systems. The level of detail ranges here from pure functional execution to cycle-accurate simulations. However, all of them are implemented single-threaded and therefore suffer from a drop in simulation performance with increasing number of cores. The gap between the performance of the simulated architecture and the host simulation performance is continuously increasing.

There are also a couple of parallel multi-core simulators available. Some of the earliest are the Wisconsin Wind Tunnel I (WWT) [14] and the Wisconsin Wind Tunnel II (WWT II) [15]. These two make also use of direct execution, where most of the code to be simulated is executed on the native hardware, and parallel discrete-event simulation. However, the WWT is dedicated to simulate a single machine only (CM-5) and both do not model the properties of heterogeneous core types and the interconnection network. For instance HORNET [16], which is basically a cycle-accurate simulator for tile-based MPSoC architectures simulating the tiles in parallel. The input to the routers comes from traffic generators, however, it is also possible to couple cycle-accurate CPU simulators or instrumented applications to the routers. As described in Chap. 3, cycle-accurate simulation is prohibitive for large scale multi-core simulation.

The authors of the full-system simulator COTSon [17] follow a different philosophy. They employ sampling techniques, which consider only interesting or representative execution phases of an applications for detailed simulation and the other phases are just simulated functionally. The simulation environment divides functional and timing simulation by defining a clear interface between both. In [18], Monchiero et al. extended the approach of COTson by splitting up the sequential instruction stream coming from the functional simulator into separate instruction streams for the target CPUs, which are simulated by separate threads. Although this approach is similar to our proposed parallelization techniques, they synchronize at the level of single instructions, whereas we only synchronize at the level of certain application events (e.g., entering or leaving atomic sections, creating or terminating new activities).

Graphite [19] is a parallel simulator based on dynamic-binary translation by PIN [20]. It makes also use of direct execution and allows modeling tiled architectures including processors, cache, memory, and a NoC. However, it is restricted to homogeneous architectures and only one core per target tile. Sniper [21] extends Graphite by interval simulation and improved accuracy of the simulator by adding a shared multi-level cache hierarchy and a cache coherency protocol. Reported are simulation speeds of 2 MIPS on an 8-core SMP machine. In PCantorSim [22], the authors improved the simulation speed of Sniper by applying fractal-based sampling techniques, achieving a speedup of up to 20 with respect to single-thread simulation with an average prediction error of 5.3%. Our parallel simulation approach targets much higher simulation speeds by further abstracting from micro-architectural details by applying a simulation approach that combines direct execution and analytical models.

5.4 Summary

In this chapter, we presented parallelization techniques for our direct-execution simulation approach for the efficient simulation of heterogeneous tile-based multi-core architectures. Proposed were four novel parallel discrete-event simulation techniques, which map activity-level parallelism within the applications to core-level parallelism on the target architecture and back to thread-level parallelism on the host machine. Experiments with parallel real-world applications have been used to compare the different techniques against each other. On a 12-core Intel Xeon processor, an average speedup of 8.2 compared to sequential simulation could be achieved for the entire set of evaluated parallel X10 applications. A maximum simulation performance of 86,000 MIPS have been achieved. The presented parallelization techniques allow for high performance simulation of complex architectures executing highly parallel workloads and thus, enable fast architecture evaluations and design space exploration. We presented case studies on the fast evaluation of the architectural design space of invasive architectures in terms number of tiles, tile organization, and processor selection within each tile. We compared different cost and performance trade-offs by evaluating different design alternatives against invasive application scenarios.

References

1. S. Roloff et al., Execution-driven parallel simulation of PGAS applications on heterogeneous tiled architectures, in *Proceedings of the 52nd ACM/EDAC/IEEE Design Automation Conference (DAC)*, San Francisco, CA, USA. HiPEAC Paper Award, 7–11 June 2015 (ACM, 2015), pp. 44:1–44:6. ISBN: 978-1-4503-3520-1. https://doi.org/10.1145/2744769.2744840
2. S. Roloff, F. Hannig, J. Teich, Fast architecture evaluation of heterogeneous MPSoCs by host-compiled simulation, in *Proceedings of the 15th International Workshop on Software and Compilers for Embedded Systems (SCOPES)*, St. Goar, Germany, 15–20 May 2017 (ACM, 2012), pp. 52–61. ISBN: 978-1-4503-1336-0. https://doi.org/10.1145/2236576.2236582
3. R.M. Fujimoto, Parallel discrete event simulation. Commun. ACM **33**(10), 30–53 (1990). ISSN: 0001-0782. https://doi.org/10.1145/84537.84545
4. D. Schafhauser, Parallel simulation of highly parallel X10 applications on heterogeneous MPSoCs. Bachelor Thesis at the Friedrich-Alexander University Erlangen-Nürnberg, Department of Computer Science 12 (Hardware/Software Co-Design). 4 Feb 2015
5. J. Teich, Invasive algorithms and architectures. IT Inf. Technol. **50**(5), 300–310 (2008). ISSN: 1611-2776. https://doi.org/10.1524/itit.2008.0499
6. J. Teich et al., Invasive computing: an overview, in *Multiprocessor System-on-Chip: Hardware Design and Tool Integration*, ed. by M. Hübner, J. Becker (Springer, Berlin, 2011), pp. 241–268. ISBN: 978-1-4419-6459-5. https://doi.org/10.1007/978-1-4419-6460-1. (Chap. 11)
7. J. Heisswolf et al., The invasive network on chip-A multi-objective many-core communication infrastructure, in *Workshop Proceedings of the 27th International Conference on Architecture of Computing Systems (ARCS)*, Lübeck, Germany, 25–28 Feb 2014 (VDE Verlag, 2014), pp. 1–8. ISBN: 978-3-8007-3579-2
8. F. Bellard, QEMU, a fast and portable dynamic translator, in *Proceedings of the Annual Conference on USENIX Annual Technical Conference (ATEC)* (2005), p. 41
9. P.S. Magnusson et al., Simics: a full system simulation platform. Computer **35**(2), 50–58 (2002). ISSN: 0018-9162. https://doi.org/10.1109/2.982916

10. T. Austin, E. Larson, D. Ernst, SimpleScalar: an infrastructure for computer system modeling. Computer **35**(2), 59–67 (2002). ISSN: 0018-9162. https://doi.org/10.1109/2.982917
11. B. Barnes, J. Slice, SimNow: a fast and functionally accurate AMD x86-64 system simulator, in *Tutorial at the IEEE International Symposium on Workload Characterization (IISWC)* Austin, TX, USA. 6–8 Oct 2005
12. M.T. Yourst, PTLsim: a cycle accurate full system x86-64 microarchitectural simulator, in *Proceedings of the IEEE International Symposium on Performance Analysis of Systems & Software (ISPASS)*, San Jose, CA, USA, 25–27 Apr 2007 (IEEE, 2007), pp. 23–34. ISBN: 1-4244-1081-9. https://doi.org/10.1109/ISPASS.2007.363733
13. N.L. Blinkert et al., The M5 simulator: modeling networked systems. IEEE Micro **26**(4), 52–60 (2006). ISSN: 0272-1732. https://doi.org/10.1109/MM.2006.82
14. S.K. Reinhardt et al., The Wisconsin Wind Tunnel: virtual prototyping of parallel computers, in *Proceedings of the ACM SIGMETRICS Conference on Measurement and Modeling of Computer Systems (SIGMETRICS)*, Santa Clara, CA, USA, 10–14 May 1993 (ACM, 1993), pp. 48–60. ISBN: 0-89791-580-1. https://doi.org/10.1145/166955.166979
15. S.S. Mukherjee et al., Wisconsin Wind Tunnel II: a fast, portable parallel architecture simulator. IEEE Concurr. **8**(4), 12–20 (2000). ISSN: 1092-3063. https://doi.org/10.1109/4434.895100
16. M. Lis et al., Scalable, accurate multicore simulation in the 1000-core era, in *Proceedings of the International Symposium on Performance Analysis of Systems and Software (ISPASS)* (2011), pp. 175–185
17. E. Argollo et al., COTSon: infrastructure for full system simulation. ACM SIGOPS Oper. Syst. Rev. **43**(1), 52–61 (2009)
18. M. Monchiero et al., How to simulate 1000 cores. SIGARCH Comput. Archit. News **37**(2), 10–19 (2009)
19. J.E. Miller et al., Graphite: a distributed parallel simulator for multicores, in *Proceedings of the IEEE International Symposium on High Performance Computer Architecture (HPCA)* (2010), pp. 1–12
20. C.-K. Luk et al., Pin: building customized program analysis tools with dynamic instrumentation, in *Proceedings of the ACM SIGPLAN Conference on Programming Language Design and Implementation (PLDI)* (2005), pp. 190–200
21. T.E. Carlson, W. Heirman, L. Eeckhout, Sniper: exploring the level of abstraction for scalable and accurate parallel multi-core simulation, in *Proceedings of the International Conference for High Performance Computing, Networking, Storage and Analysis (SC)* (2011), pp. 52:1–52:12
22. C. Jiang et al., PCantorSim: accelerating parallel architecture simulation through fractal-based sampling. ACM Trans. Archit. Code Optim. **10**(4), 49:1–49:24 (2013)

Chapter 6
ActorX10 and Run-Time Application Embedding

The last chapters were dedicated to the efficient, flexible, yet timing-approximate simulation of heterogeneous MPSoCs of multiple parallel invasive X10 applications. The presented techniques have finally been integrated into the full-system simulator InvadeSIM and pave the way for detailed simulation-based studies on invasive programming concepts, application mapping approaches, and timing-predictability through invasive computing principles. This becomes especially important when multiple applications from different domains are executed on the same MPSoC and expected to guarantee certain user requirements regarding timing, reliability, or security. Examples of such applications are real-time or safety-critical parallel applications for embedded systems from the domain of mobile and multimedia communication, robotics, automotive, or avionics. Particularly streaming applications from signal, audio, and image processing have soft or even hard real-time requirements. Highly-parallel MPSoCs provide a huge amount of computing power to accomplish the tasks, but due to sharing of resources, e.g., when applications access on-chip communication and memory, applications suffer from the interference with other applications, uncertainty of dynamic workloads, and state of the hardware. In order to tackle these challenges and to guarantee real-time requirements at design time independent from dynamic workloads, so-called *hybrid application mapping* (HAM) approaches [1–3] have been introduced and are studied in invasive computing. Such approaches characterize applications at design time either based on timing analysis [4] or on simulation to automatically generate a set of resource constraints on claims that when provided at run time will satisfy the real-time requirements. At run time, applications are embedded on an MPSoC depending on the currently available resources. The dynamic resource reservation concepts of invasive computing are the key enabler for such HAM approaches, since they provide the required isolation of resources allocated to each application, which allows for composable[1] and

[1]The concept of composability has been introduced in [5] and states that a composable design allows analyzing execution properties of an application already at design time without knowing the other applications. This significantly reduces the effort of verifying real-time requirements of

S. Roloff et al., *Modeling and Simulation of Invasive Applications and Architectures*, Computer Architecture and Design Methodologies,
https://doi.org/10.1007/978-981-13-8387-8_6

timing-predictable application executions. In the context of these HAM approaches, the contributions of this chapter are (a) the design and development of a novel *actor-oriented programming library* called ActorX10, which allows describing computation and communication of stream processing applications in X10 formally being the fundamental application model used in HAM, (b) the development and evaluation of an efficient mapping algorithm based on backtracking to find constellations of resources fulfilling the statically determined resource constraints at run time, and c) two simulation-based real-world case studies from the computer vision domain to demonstrate the capabilities of HAM to provide predictable execution of stream processing applications on heterogeneous MPSoCs and to guarantee statically analyzed best and worst-case timing requirements on latency and throughput. The presented concepts and implementations have been published in the following papers. General definitions about actor-oriented programming, the implementation of ActorX10 as well as two programming case studies have been presented in [6]. The run-time mapping algorithm based on backtracking including an experimental evaluation are published in [7]. Two simulation-based case studies evaluating the predictable execution properties of invasive applications have been presented in [1, 8].

6.1 Fundamental Definitions

Before the details of our contributions are explained, some fundamental definitions on predictability, isolation, and the HAM approach from [1] are introduced.

6.1.1 *Definition of *-Predictability*

In the realm of embedded real-time systems, predictability is usually associated with the aspect of timing. For example, a system is called timing predictable, if it is possible to provide an upper bound on the system's reaction time to the occurrence of a certain event.[2] Timing predictability inherently deals with some level of *uncertainty*, which may arise from assumptions on the *input* to a program (e.g., an image processing algorithm takes longer to process 4K images than images of VGA resolution) or the *environment* in which the program will be executed (e.g., the state of the caches or the pipeline of a processor or the busses connecting the processors). According to the definitions in [9], the notion of predictability should capture if and to what level of precision a specified property can be predicted. Figure 6.1 illustrates the introduced terms. Let o denote the objective or aspect of a program to be predicted,

an application as not all possible run-time scenarios have to be considered. In invasive computing, composability is established through the exclusive reservation of resources to each application (isolation on demand).

[2]As an example, the airbag controller in a car must react upon the event of an abruptly negative acceleration at a crash within a few microseconds and fire the airbags as a reaction to this event.

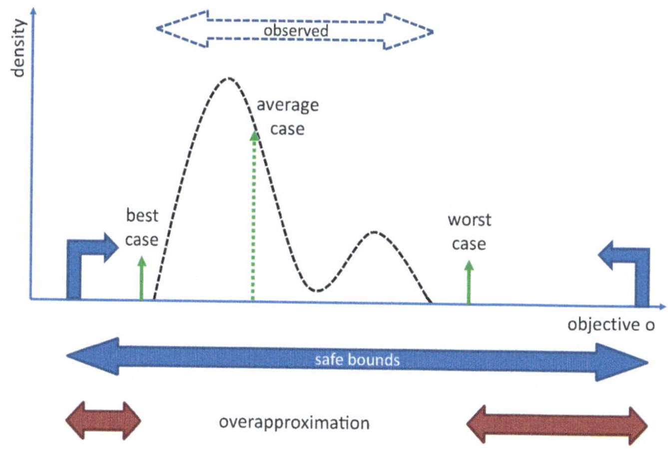

Fig. 6.1 Predictability in terms of observed and analyzed bounds of an objective or aspect o of a program [1]

then different observations of this property lead to a distribution with an observed best case (assuming the objective o to be minimized) and an observed worst case as well as bounds determined by an analysis technique. These are called *safe* or sometimes also *sound* in the sense that the upper bound is always bigger than any observable bound and the lower bound is always lower than any observable bound.

In the literature, different definitions of predictability exist. However, they have some shortcomings and are not able to characterize absolute bounds on timing. For example, the definition according to Grund et al. [10] does not indicate the variation of the property under concern. Another definition by Thiele and Wilhelm [11] characterizes the analysis technique rather than system and program properties. According to Henzinger [12], timing predictability should be an inherent system property, which greatly sacrifices system efficiency, because all programs are required to hold this property. Thus, Teich et al. [1] introduced a new definition to predictability characterizing it by *predictability intervals and distributions*. Furthermore, this approach is generalized to consider not only timing aspects, but also other non-functional properties such as power or costs, because they might also be subject to bounds during the execution of a program.[3] This generalization called **-predictability* is defined as follows:

Definition 6.1 ([1]) Let o denote a non-functional property of a program (implementation) p and the uncertainty of its input (space) given by I and environment by Q. The *predictability (marker)* of objective o for program p is described by the interval

$$o(p, Q, I) = [INF_o(p, Q, I), \ldots, SUP_o(p, Q, I)] \tag{6.1}$$

[3]An example is battery-driven embedded devices such as a mobile phone that must never consume more than 5 W maximum power due to thermal problems that might arise otherwise.

where INF and SUP denote the infimum and supremum under variation of state $q \in Q$ and input $i \in I$, respectively.

As an example, a program p that under the uncertainty of dynamic power management and input leads to a total power consumption P between 1 and 2 W is described by $P(p, Q, I) = [1, \ldots, 2]$. Moreover, a variation within a predictability marker can be annotated with a distribution, e.g., a uniform or any other discrete or continuous distribution.

6.1.2 Restriction and Isolation

In the previous section, the notion of *-predictability has been introduced as the variability of any non-functional execution property o of a program p under the uncertainty as observable through variation of input space I and state space Q program p is executed. In this section, the two concepts of *restriction* and *isolation* are explained to reduce the predictability intervals of programs on multi-core machines. *Restriction* describes a set of methods to reduce the input space I to a substantially smaller input space I', while *isolation* describes methods that are able to reduce the state space Q that influences the quality o and its variability.

Definition 6.2 ([1]) A *restriction* denotes a subset I' of the input space I a program (implementation) may experience, i.e., $I' \subseteq I$.

Here is a list of examples that may improve *-predictability through restriction:

- Maximal size of a matrix, vector, or image resolution of an image filter to be processed (latency, memory requirements),
- Environmental restrictions: Temperature, radiation (reliability), attacker model (security), and
- Approximate computing.

Definition 6.3 ([1]) An *isolation* denotes a subset Q' of the state space Q a program (implementation) may experience, i.e., $Q' \subseteq Q$.

Examples of how to achieve isolation are:

- Simpler cores,
- Sophisticated co-analysis of architecture and program execution,
- Resource reservation protocols, e.g., TDMA, cache-protocols,
- Virtualization, and
- No sharing of resources, e.g., using *invasive computing*.

Figure 6.2 shows how restriction and isolation together may improve the predictability of a program p. In Sect. 6.4, two simulation-based case studies of a complex soft real-time computer vision application are presented showing how invasive computing allows isolating applications from each other spatially or temporally and therefore improving predictability by reducing latency and throughput jitter on NoC-based heterogeneous MPSoC architectures.

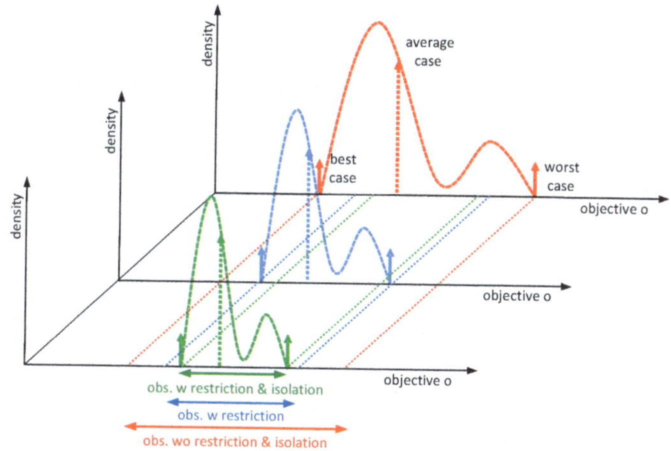

Fig. 6.2 Visualization of the effects of restriction and isolation on the predictability of an objective o [1]

Listing 6.1 Examples of requirements in InvadeX10 implemented by specific parametrizable X10 annotations [1].

```
/* Examples of performance requirements */
@REQUIRE(Latency(0, 65, "ms", "soft"))
@REQUIRE(Throughput(25, 100, "fps", "hard"))
/* Example of a reliability requirement */
@REQUIRE(PFH(0.001, 0.0000001))
/* Example of a power requirement */
@REQUIRE(Power(1, 2, "W", "soft"))
```

6.1.3 *Requirements Versus Constraints*

It might be difficult or even impossible for a programmer to specify by hand constraints on number and type of resources to achieve a desired quality of program execution. Thus, invasive computing allows the programmer specifying so-called requirements on objectives of execution instead. Listing 6.1 gives an example of a class of *requirements* that may precede a program segment. These requirements, such as, e.g., a timing requirement of a maximal latency bound of 65 ms, or a throughput requirement of minimally 25 frames/s *frames per second* (FPS) will be analyzed at compile time. A *hybrid mapping methodology* that explores constellations of claims for a given MPSoC platform that will satisfy these requirements will be described in the next section.

For example, according to Definition 6.1, a power requirement, as shown above, is satisfied for an *i*-let p or code segment therein, if the constraint set of a suitable claim of resources will guarantee $INF_P(p, Q, I) \geq 1$ and $SUP_P(p, Q, I) \leq 2$. The difference between hard and soft requirements is that hard requirements must never be violated for whatever reason, whereas soft requirements shall be satisfied most of the time, but their occasional violation may be tolerated. The code listing example also demonstrates a reliability requirement in terms of bounds on tolerable *probabilities of failure per hour* (PFHs) as well as a soft power consumption requirement of the resources claimed. The power requirement will be translated into *constraints* on the power management of the claim, e.g., a restriction of allowed DVFS settings on the claimed cores.

6.1.4 Hybrid Application Mapping

To fulfill the aforementioned user requirements for applications, admitted at run time, the following design flow is used. It automatically determines claim constraints that adhere to a set of given requirements and then replaces the requirement annotations in the X10 source code by the respective set of claim constraints. Timing predictability is achieved by a HAM approach, which combines design-time application characterization with run-time application mapping, see Fig. 6.3. This separation is enabled by the exclusive resource reservation and isolation in invasive computing. It allows analyzing applications independently of each other and therefore, determining resource allocations fulfilling the requirements already at design time, which in turn drastically reduces the complexity to find a feasible solution at run time. Applications are programmed in *ActorX10*, an X10 library of a formally specified actor model based on the PGAS principles, which is the first main contribution of this chapter and will be presented in Sect. 6.2. Formally, an actor model of an application can be described by a graph of communicating actors, each triggering the execution of a task per activation by data (tokens) on the arcs that represent data dependencies. At design time, the application is analyzed by a design space exploration with respect to a given *target platform*. Here, different allocations of computation and communication resources are selected and the tasks and messages of the application actors are mapped onto them. These so-called *operating points* are then evaluated according to various objectives (e.g., latency or energy consumption) by simulation using InvadeSIM or analytical techniques as presented in [4]. They are iteratively improved by an optimization loop and the best solutions are kept in a Pareto front. Each Pareto point is accompanied by a so-called *constraint graph* [4]. It specifies which groups of tasks (task clusters) have to be mapped onto which resource type and which communication constraints between the task clusters have to be fulfilled (e.g., hop distance or required service level along the routing path). At run time, this data structure is passed to the *run-time management* by the `invade` call. It tries to embed one of the available operating points onto the target platform depending on the current system state by looking for suitable claim constellations adhering to the constraints given by the respective constraint graph. This is formulated as a *constraint satisfaction problem* (CSP) and solved by an efficient *backtracking algorithm*, which is the second main contribution of this chapter and will be presented in Sect. 6.3. If

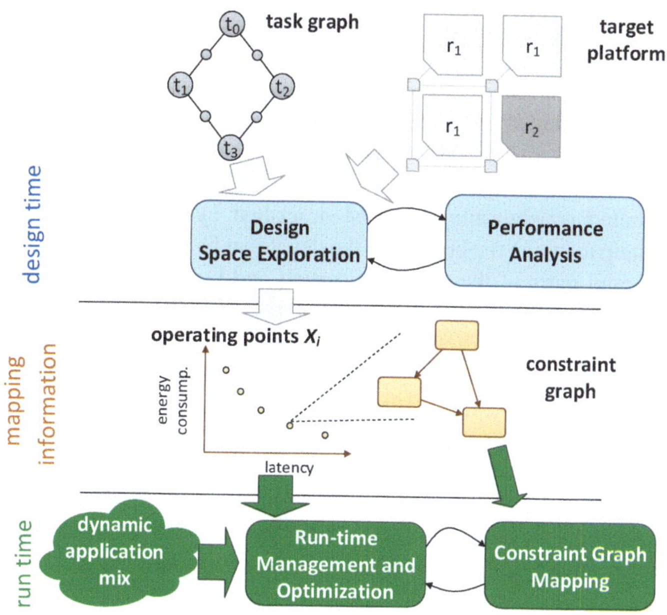

Fig. 6.3 Overview of the introduced hybrid application mapping approach [4]

found and returned, the application will meet the execution requirements within the statically analyzed intervals of uncertainty.

6.2 ActorX10

As explained in the last section, the introduced HAM approach relies on programs described in X10. However, for analyzability and mapping of streaming applications, dataflow models of computation are more suitable than standard X10. Here, actor specifications that are described by periodically executing programs triggered by the availability of data, and distributed over heterogeneous tiles of processors, are modeled. For allowing a formal analysis of such actors, we propose *ActorX10*, a novel actor-oriented programming library based on X10, which enables to formally implement actor functionality and its communication based on the APGAS principles. The APGAS programming model is a powerful computing paradigm for multi-core and massively parallel computer architectures. It allows for the dynamic creation and distribution of thousands of threads amongst hundreds of nodes in a cluster computer within a single application. For programs of such complexity, communication patterns between threads concurrently running on different places is not always obvious and appropriate higher level abstractions on computation and communication are necessary for performance analysis and optimization. Here, actor-based programming may provide the right abstraction for a wide range of applications. The realized actor model explicitly exposes communication and decouples these from the control flow of the concurrently executed application components. Its capabilities and advantages are introduced and demonstrated for two applications from the embedded system and HPC domain, i.e., an object detection chain and a proxy application for the simulation of tsunami events.

Actor-oriented programming, as first described by [13] and later formalized by [14], can help to lessen the burden on the developer by making the communication between different parts of the program explicit and by separating the control flow aspects from the computational aspects of a program. *Actors* also most naturally describe concurrently, periodically executed parts of computations and communicate by sending and receiving data using *ports* and *channels*. Furthermore, their activation is data-driven which enables to execute as soon as data is available in a self-scheduled manner. Moreover, in contrast to the conventional X10 *place shift* by the keyword `at`, which implicitly captures, serializes, and deserializes data, the actor model explicitly models the data that will be sent. Finally, if developers formalize their computation in this way, they are able to analyze their streaming applications using design space exploration, where different actor-to-place mappings can be evaluated using a graph-based approach as presented in Sect. 6.3.

In the following, we will introduce an X10 actor library that maps the principles of actor-oriented programming onto the APGAS paradigm. The library uses the features provided by the APGAS environment to implement its functionality as well as to enable features such as migrating actors between different places. Our contributions may be summarized as follows:

- Definition of a formalism that incorporates the APGAS paradigm into an actor model,
- Implementation of the formalism by an actor library in X10, and
- Demonstration of the viability of combining the actor model with X10 using examples from two different application areas, namely a simulation of the shallow water equations from the domain of HPC and a computer vision application from the area of embedded systems.

6.2.1 Actor Model

In this section, we formally define the notions of actors and actor graphs. This general model is the basic formal underpinning of the proposed X10 actor library and—as we will discuss later—can be specialized to adhere to *different models of computation* (MoCs).

Definition 6.4 (*Actor graph* [15]) An actor graph is a directed graph $G_a = (A, C)$ containing a set of actors A and a set of channels $C \subseteq A.O \times A.I$ connecting actor output ports $A.O$ with actor input ports $A.I$. Each channel has a buffer size n :

$C \rightarrow \mathbb{N}^+$, and a possibly empty sequence of initial tokens denoted by $d : C \rightarrow D^*$ where D^* denotes the set of all possible finite sequences of tokens.

The communication between actors through channels connected by ports occurs in FIFO order. This is natural for many applications in, e.g., stream processing where all data produced must be consumed in the same order. As indicated by the formal definition, actors are only permitted to communicate with each other via channels and there is no *implicit* communication allowed, e.g., via shared data structures. An actor is thereby constrained only to consume data items also called *tokens* from channels connected to its input ports and to produce tokens on channels connected to its output ports. It is also possible to place initial tokens in FIFO channels, e.g., $d(c) = \langle 1, 2 \rangle$ to denote that channel c is initialized with two initial tokens with the values 1 and 2, respectively.

In actor-based modeling, rules must be specified how and when actors may be activated to perform a computation. Thereby, it must be also specified, how many tokens are read and consumed from input ports, and how many tokens will be produced on which output ports during such a transition also called *firing*. In our model of an actor, the communication and firing behavior is strictly separated from the functions that are performed on the data at input ports. Moreover, these firing rules can be formally described by a finite state machine as follows.

Definition 6.5 (*Actor* [15]) An *actor* is a tuple $a = (I, O, F, R)$ containing actor ports partitioned into a set of actor input ports I and a set of actor output ports O, a set of functions F, and a finite state machine R called *firing* finite-state machine *(FSM)*. The functions encapsulated in an actor are partitioned into so-called *actions* $F_{\text{actions}} \subseteq F$ and *guards* $F_{\text{guards}} \subseteq F$. Functions are activated during a so-called *firing transition* of the FSM R, which unambiguously formalizes the communication behavior of the actor (i.e., the number of tokens consumed and produced in each actor firing). Actions may produce results in the form of output tokens residing in the FIFO channels connected to the actor output ports. Using guards, more complex models of computation may be modeled. In particular, the activation of actors is based not only on the availability of a minimal number of tokens on the input ports, but also on their values. Guards return a Boolean value and may be assigned to each transition of the FSM of an actor.

Definition 6.6 (*Actor (Firing) FSM* [15]) The *firing FSM* of an actor $a \in A$ is a tuple $R = (Q, q_0, T)$ containing a finite set of states Q, an initial state $q_0 \in Q$, and a finite set of transitions T. Moreover, a *transition* of an FSM is a tuple $t = (q, k, f, q') \in T$ contains the current state $q \in Q$, an activation pattern k, the respective action $f \in a.F_{\text{actions}}$, and the next firing state $q' \in Q$. The activation pattern k is a Boolean function which decides if transition t can be taken (*true*) or not (*false*) based on: (1) a minimum number of available tokens on the input ports $a.I$, (2) a set of guard functions $F' \subset F_{\text{guards}}$, and (3) a minimum number of free places in the channels connected to respective output ports.

An example of a simple actor for computing the square root of a number and its firing FSM is shown in Fig. 6.4. In the initial state S_1 of the firing FSM, the actor

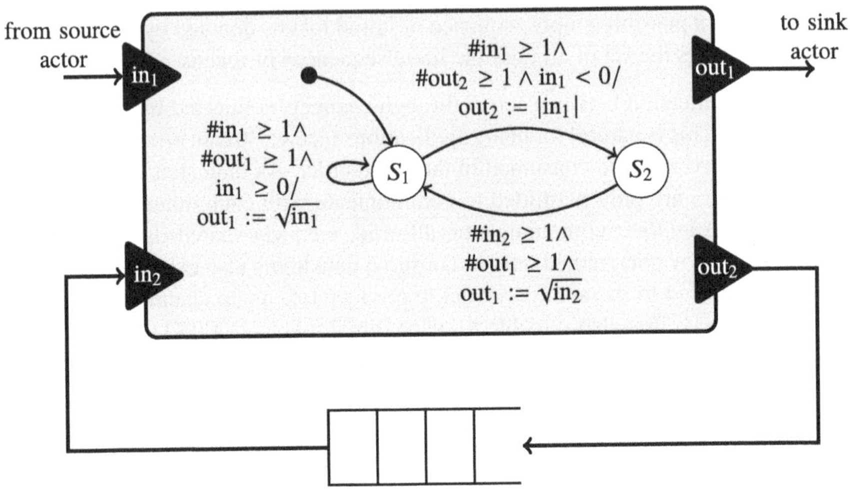

Fig. 6.4 Example from [15] of an actor with two input and two output ports calculating the square root of a number. In the initial state S_1 of the firing FSM, the actor waits for one token on its input port in_1 and for free space of one token on its output port out_1. If the number is positive, the square root is computed and the result token is produced on output port out_1. However, if the number is negative, its absolute value is calculated and produced on output port out_2 which is fed back over a channel to input port in_2. In state S_2, the actor waits and consumes a token on input port in_2 and then advances back to the initial state S_1

waits for at least one token on its input port in_1 ($\#in_1 \geq 1$) as well as for free space of at least one token on its output port out_1 ($\#out_1 \geq 1$). If the value of the first arriving token is greater than zero (guard: $in_1 \geq 0$), then the square root result is computed (action: $\sqrt{in_1}$) and the result token produced directly on output port out_1 ($out_1 := \sqrt{in_1}$). However, if the number is negative, its absolute value is calculated first and produced as a token on output port out_2 which is fed back over a channel to input port in_2. In state S_2, the actor waits for its activation by at least one token on this input port in_2 as well as for free space of one token on the output port out_1 and then advances back to the initial state S_1 once having computed the square root of this token on input in_2.

The concept of actor FSM is an extension of the state machines of the *FunState* model of computation [16, 17] standing for functions driven by state machines, by additionally providing mechanisms to check the availability of sufficient space on output channels before a transition can be taken. The state transitions of an actor FSM are annotated each with a so-called *activation pattern* which is a Boolean expression.

6.2.1.1 Mapping the Actor Model onto the PGAS Model

The aforementioned formal model separates control flow and data flow. It encapsulates data processing in side-effect functions called *actions*.

The X10 PGAS model introduces the notion of places and giving us now the opportunity to distribute an actor graph among the available places in order to exploit processing power of the underlying cores and for optimal workload distribution. Informally, each actor and each channel must be assigned a certain place on which they reside during the execution of the actor graph. For static actor graphs, an optimal actor and channel distribution with respect to different objectives, such as performance, power efficiency, or temperature distribution, can be found or at least approximated by applying techniques of design space exploration [3]. In this chapter, we consider a manual distribution of actors to places. In case of more adaptive scenarios, even the dynamic migration of actors and channels to other places at run time may be considered. Techniques for static design space exploration [3] and dynamic place assignment are ongoing work.

In this chapter, we present a first step into that direction by providing an implementation of the actor model in X10 as a library called ActorX10 [6]. Further, we will show how the formal actor model can be implemented by means of available X10 language constructs. Then, it is shown that different types of applications can be realized using the actor library. Two concrete examples from the multimedia and the high-performance computing domains are subsequently presented.

6.2.2 X10 Implementation

In this section, we present the basic data structures and primitives of our library-based implementation of the previously introduced actor model and semantics in X10.

Actor

According to the definition of an actor (see Sect. 6.2.1), we introduce the fundamental abstract class `Actor`. Each `Actor` consists of two lists, one for its input (I) and output (O) ports, and provides factory methods for creating ports. `Actor` defines an abstract method `act`, which has to be implemented by its subclasses and contains the (firing) state machine (R) as well as all functions (F) of the actor. Each actor and port possesses a unique name.

An actor is defined by creating a subclass of the `Actor` class and implementing the method `act`. The new subclass may define its own constructors. Ports are created using the factory methods of the class `Actor` and are used within the `act` method to send tokens to and receive tokens from other actors.

Port

The input and output ports of an actor define its logical interface to the outside world. From these, the application developer may deduce the type of data that is produced on, respectively consumed on a port of an actor. Accordingly, we introduce the generic classes `InPort[T]` and `OutPort[T]`. The type parameter `T` allows for a type-safe communication between actors due to the fact that only ports with the same type can be connected by a channel. An actor implementation may `read` data from an `InPort` and `write` data to an `OutPort`.[4] These communication primitives fail if no data is available or no space is left at the corresponding channel. Methods are provided that allow to check these activation patterns non-destructively (e.g., `in1(1)` checks for one input token on the input port `in1`). If ports are accessed despite not being ready, an exception is thrown. Internally, a port has a global reference to a `Channel` and forwards all read and write requests to the corresponding X10 object on its respective home place, which may include a place shift. These channel references are set in the port mapping phase during the creation of an actor graph.

Example of an X10 Actor

An implementation of the square root example actor from Fig. 6.4 is depicted in Listing 6.2. The class `SquareRootActor` is derived from the `Actor` class and two input and two output ports of type `double` are created. A state variable—initialized with state S_1—indicates the current state of the actor. The implementation of the state machine is described by the body of the `act` method. In state S_1, the two activation conditions emanating state S_1 are checked. According to line 11, at least one token must be present on input port in_1. The guard method `peek` in line 11 tests the value of the first token in the FIFO channel connected to port in_1 non-destructively.[5]

Channel

A channel is implemented by the generic class `Channel[T]`. It provides a FIFO implementation of limited capacity. Similar to the port classes, it has a type parameter `T` that matches the type of its writing and reading port. Data of type `T` is stored in a thread-safe queue which allows concurrent read and write accesses. Furthermore, `peek` and `poke` methods allow the caller to check the presence or absence of tokens in the queue. As this communication pattern may involve several places, each token must be *deeply copied* from the output port place via the channel place to the input port place, if these places are not the same. Therefore, a typical communication path between two actors looks like this: $\texttt{Actor}_{A} \rightarrow \texttt{OutPort[T]}_{A} \rightarrow \texttt{Channel[T]} \rightarrow \texttt{InPort[T]}_{B} \rightarrow \texttt{Actor}_{B}$.

Actor Graph

The class `ActorGraph` holds lists of global references to `Actors` and `Channels`, according to the definition of actor graphs. Constructing an actor graph involves

---

[4] In a certain state of a firing FSM, the activation patterns are evaluated. Yet, tokens on an input port are only destructively consumed by `read` once an activated transition fires that required a certain number of tokens on that port for activation. Once a transition is selected to fire, the respective number of tokens are consumed. Whereas, the transition completes by producing output tokens only once the actions as evoked during the chosen transition have finished.

[5] Note that whereas our formal actor model introduced in Sect. 6.2.1 assumes a non-deterministic choice of transition in case multiple transitions should become simultaneously activated, our X10 implementation implements a priority given by the order in which the code checks the activation conditions.

Listing 6.2 Square root actor from Fig. 6.4 using ActorX10

```
public class SquareRootActor extends Actor {
    val in1 = new InPort[double]("in1");
    val in2 = new InPort[double]("in2");
    val out1 = new OutPort[double]("out1");
    val out2 = new OutPort[double]("out2");
    var state : int = 1;
    public def this() { super("SquareRoot"); }
    public def act() {
        switch (state) {
            case 1: /* S1 */
                if (in1(1) && out1(1) && in1.peek() >= 0) {
                    val sqrtNumber = Math.sqrt(in1.read());
                    out1.write(sqrtNumber);
                    state = 1;
                } else if (in1(1) && out2(1) && in1.peek() < 0) {
                    val absNumber = Math.abs(in1.read());
                    out2.write(absNumber);
                    state = 2;
                }
                break;
            case 2: /* S2 */
                if (in2(1) && out1(1)) {
                    val sqrtNumber = Math.sqrt(in2.read());
                    out1.write(sqrtNumber);
                    state = 1;
                }
                break;
        }
    }
}
```

adding actors (`addActor`) and connecting input and output ports (`connectPorts`) of these actors via channels. A connection between two ports is established by creating a `Channel` objects and exchanging references between port and channel on both sides. Actors can be freely migrated to other places by calling the `moveActor` method prior to the start of the actor graph's execution. During migration, the state of the actor is copied to the new place and all affected channel and port references are updated. This method transfers the actor state to another place and ensures all port and channel references of connected actors point to the copied actor on the new place instead of the original. The actor graph may be started by calling the `start` method, which internally calls the `start` method of all its actors on their respective place. This method only returns once all actors are finished.

Example of an Actor Graph Creation

An example of an actor graph creation using our library is depicted in Fig. 6.3. First, all relevant actors are instantiated using their individual constructors (line 2–4). In this

Listing 6.3 Actor graph generation and execution of the square root example. The three actors `SourceActor`, `SinkActor`, and `SquareRootActor` are connected with each other properly. All actors are mapped to the place `here` except the actor `SquareRootActor` itself, which is moved to the place `here.next()` according to the code above. The method `start` then initializes all actors to their respective initial state

```
1 public static def main(args:Rail[String]) {
2     val source = new SourceActor(100);
3     val sink = new SinkActor();
4     val squareRoot = new SquareRootActor();
5     val ag = new ActorGraph();
6     ag.addActor(source);
7     ag.addActor(sink);
8     ag.addActor(squareRoot);
9     ag.connectPorts(source.outPort, squareRoot.in1);
10    ag.connectPorts(squareRoot.out1, sink.inPort);
11    ag.connectPorts(squareRoot.out2, squareRoot.in2);
12    ag.moveActor(squareRoot, here.next());
13    ag.start();
14 }
```

case, the source actor generates a list of 100 random numbers. After instantiating the actor graph in line 5, all actors are added to it and their ports are properly connected. All actors are mapped to the place `here` except the actor `SquareRootActor` itself which is moved to the place `here.next()`. Finally, the whole actor graph is started in line 13.

6.2.3 *Real-World Examples Modeled in ActorX10*

In this section, we showcase two complex real-world X10 applications that were adapted to use our actor library.

6.2.3.1 Object Detection

Object detection is a typical computer vision task, for example as used in the robotics domain. Here, a previously trained object is detected in a stream of images. There are several approaches to solving this problem, most of which involve chaining together multiple algorithms. Our implementation uses a *scale-invariant feature transform* (SIFT)-based approach and comprises of the following sub-algorithms:

1. Harris operator to detect corners in an image. These act as a basis for the features of the image, which are later used to compare against the object features.

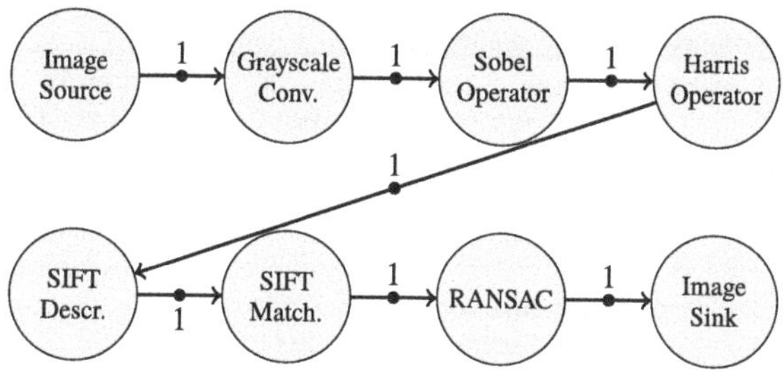

Fig. 6.5 An object detection algorithm chain in SDF graph semantics [6]

2. SIFT description to transform the corners of the image into 128-dimensional feature descriptors. The descriptors are scale and rotation-invariant, which allow detection of an object, even if it is rotated and/or scaled.
3. SIFT matching to compare the current image features with the object features. Here, the goal is to find for each image feature the nearest neighbor out of the object features in the 128-dimensional space. Specialized data structures such as k-d trees are used to handle this highly-dimensional search space.
4. *Random sample consensus* (RANSAC) algorithm to register the matched features to the shape of the actual object. This algorithm eliminates outliers and detects the objects.

All actors are executed periodically and data is exchanged from one stage to the next. This algorithm chain intuitively maps to the actor model where each algorithm is encapsulated into an actor. The representation of the full chain in *synchronous data flow* (SDF) graph semantics is depicted in Fig. 6.5 and an implementation of the SIFT matching actor using our actor library is shown in Listing 6.4.

The actors only exchange a single token with their neighbors. It triggers the firing of the next actor. In this context, tokens can be whole images, derivatives from images, lists of corners and descriptors as well as coordinates. Due to data dependencies, all steps in the computation of one frame have to be executed in sequence. However, actorized modeling allows for the exploitation of pipeline parallelism and hence for the concurrent processing of several stages of the algorithm chain. Using the actor library, the throughput of this application for a stream of input images could be substantially improved. The latency for one image, of cause, remains the same. However, the slowest link in the chain limits the maximum throughput. Parallel X10 constructs can aid in the in the expedition of slow actors as well as appropriate mappings of algorithms to places containing an accelerator. Evaluating different mapping decisions can be done in the context of a DSE.

6.2.3.2 Shallow Water Equations (SWE-X10)

In [18], Pöppl presented a solver for the shallow water equations in X10 (SWE-X10). An overview of the implementation of SWE-X10 in ActorX10 is provided by Pöppl in [6] and presented in this section. The shallow water equations are a set

Listing 6.4 Implementation of the SIFT matching actor

```
public class SIFTMatchingActor extends Actor {
    val inPort = new InPort[SIFTMatchingToken]("in");
    val outPort = new OutPort[RANSACToken]("out");
    val kdtree = new KdTree();
    var state : int = 1;

    public def act() {
        /* Check end condition */
        if (inPort(1) && outPort(1)) {
            if (inPort.peek().isFinished()) {
                inPort.read();
                outPort.write(new RANSACToken());
                stop();
            }
        }
        /* Execute state machine */
        switch (state) {
            case 1:
                if (inPort(1)) {
                    val token = inPort.read();
                    kdtree.initialize(token.getFeatures());
                    state = 2;
                }
                break;
            case 2:
                if (inPort(1) && outPort(1)) {
                    val inToken = inPort.read();
                    val features = inToken.getFeatures();
                    val matchedFeatures = SIFTMatching.apply(
                        kdtree, features);
                    val outToken = new RANSACToken(
                        matchedFeatures);
                    outPort.write(outToken);
                }
                break;
        }
    }
}
```

of hyperbolic *partial differential equations* (PDEs) used, for example, to simulate tsunami events. A property of hyperbolic PDEs is that changes in the simulation domain propagate locally instead of instantaneously over the entire domain. SWE-X10 exploits the benefits of the local wave propagation, e.g., only by performing computations in actors that actually exhibit non-static behavior. The solver is based on a C++ implementation first described in [19].

In both implementations, the simulation domain is discretized into a Cartesian grid of equally sized cells with constant cell data. Moreover, both are based on a finite volume time stepping scheme, where the next time step is computed based on the state of the unknowns in the previous time step. The update is computed by solving a Riemann problem at each edge between every two neighboring cells in the simulation. This dependency becomes problematic once the number of simulated cells gets large enough to warrant its distribution amongst multiple places. Now the cells at the boundary of each two neighboring distributed regions have to be exchanged between places in every time step.

To solve the problem of data exchange without global coordination, the actor model is applied. We subdivided the simulation domain into $d_y \times d_x$ quadratic patches of equal size for mapping these to tiles of resources. Each patch has a coordinate (i, j) that corresponds to its position in the simulation domain. These are each controlled by a single actor $a_{i,j} = (I_{i,j}, O_{i,j}, F_{\mathrm{SWE}}, R_{\mathrm{SWE}})$. Each of these has one outgoing and one incoming port for every patch boundary that is not an outer boundary of the simulation domain.

$$\begin{aligned} I_{i,j} = \{ip_{k,l} | \big(k \in \{i-1, i+1\} \wedge 0 \leq k < d_y \wedge l = j\big) \\ \vee (l \in \{j-1, j+1\} \wedge 0 \leq l < d_x \wedge k = i)\} \\ O_{i,j} = \{op_{k,l} | \big(k \in \{i-1, i+1\} \wedge 0 \leq k < d_y \wedge l = j\big) \\ \vee (l \in \{j-1, j+1\} \wedge 0 \leq l < d_x \wedge k = i)\} \end{aligned} \tag{6.2}$$

In our X10 implementation of the `SimulationActor` class, depicted in Listing 6.5, the `Rail inPorts` corresponds to $I_{i,j}$, while `outPorts` corresponds to $O_{i,j}$. The called functions F_{SWE} are defined as follows: `receiveData` and `sendData` read data from the incoming ports and write to the outgoing ones, respectively, `computeNumericalFluxes` determines the updates for the next timestep by solving Riemann problems for every edge between two neighboring cells, and `updateCells` updates the variables stored in the cells and progresses the simulation.

The finite state machine R_{SWE}, given in Fig. 6.6, defines the control flow of the actor. Its implementation is the `act` method of the `SimulationActor` class, shown in Listing 6.5. The method is called whenever there is an event on one of the actor's ports. There are three states the computation may embody: in the *Initial* state, the actor will write the initial cell values from the boundary of its domain to its `OutPorts`, thereafter switching to the *Compute* state. Here, the actor reads the data from the neighboring unknowns, computes a new time step, updates the block's unknowns and sends the new values from its own boundaries to the neighboring actors whenever there is data in all of the `InPorts` and capacity in all of the `OutPorts`. As long as the simulation has not progressed past `maxTime`, the actor will remain in the *Compute* state; afterwards, its state will move to *Done* and the computation concludes.

After the structure of an individual actor has been defined, a description of the actor graph can be formulated. The simulation domain is subdivided into $d_y \times d_x$ patches each controlled by one actor. Hence, the actor graph, G_a^{SWE}, may be described as follows:

Listing 6.5 Implementation of the `SimulationActor`

```
public class SimulationActor extends Actor {
    private val outPorts : Rail[OutPort[Data]];
    private val inPorts : Rail[InPort[Data]];
    private val communicators : Rail[BlockCommunicator];
    private var block : MoveablePatchSWEBlock;

    public def act() {
        if (initialSend && mayRead()) {
            sendData(currentTime, false);
            initialSend = false;
        } else if (currentTime < endTime && mayRead() && mayWrite()) {
            receiveData();
            block.computeNumericalFluxes();
            block.updateCells(dt);
            currentTime += dt;
            sendData(currentTime, (currentTime >= endTime));
        } else if (currentTime >= endTime) {
            stop();
        }
    }

    private def mayRead() =
        inPorts.reduce((i:InPort[Data], b:Boolean) => ((i == null || i(1)) &&
              b), true);

    private def mayWrite() =
        outPorts.reduce((o:OutPort[Data], b:Boolean) => ((o == null || o(1))
              && b), true);

    private def receiveData() {
        for ([i] in communicators) {
            if (communicators(i) != null) {
                val data = inPorts(i).read();
                communicators(i).setGhostLayer(data);
            }
        }
    }

    private def sendData(curTime:Float, dead:Boolean) {
        for ([i] in communicators) {
            if (communicators(i) != null) {
                val data = communicators(i).getCopyLayer(curTime, dead);
                outPorts(i).write(data);
            }
        }
    }
}
```

$$
\begin{aligned}
G_a^{\mathrm{SWE}} &= (A_{\mathrm{SWE}}, C_{\mathrm{SWE}}) \\
A_{\mathrm{SWE}} &= \left\{ a_{i,j} \mid 0 \leq i < d_y \wedge 0 \leq j < d_x \right\} \\
C_{\mathrm{SWE}} &= \left\{ (op_{k',l'}, ip_{i',j'}) \mid op_{k',l'} \in O_{i,j} \wedge ip_{i',j'} \in I_{k,l} \wedge \right. \\
&\qquad \left. (|i-k| + |j-l| = 1) \wedge i' = i \wedge j' = j \wedge k' = k \wedge l' = l \right\} \\
n_{\mathrm{SWE}}(c) &= 1 \qquad \forall c \in C_{\mathrm{SWE}}
\end{aligned}
\tag{6.3}
$$

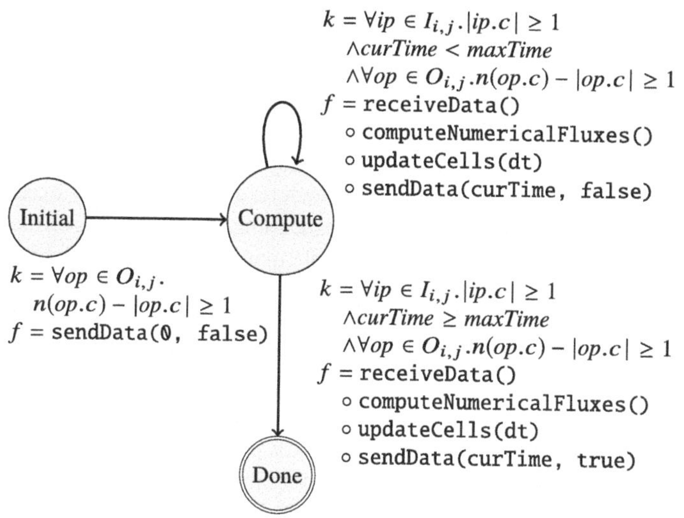

Fig. 6.6 Finite state machine for a simulation actor in SWE-X10. It has three states: In the initial state, the setup takes place, the second state represents the main loop, with each transition being the computation of one timestep. The last state is reached when the simulation is finished [6]

Fig. 6.7 SDF graph representation of a simulation run with 3 × 3 actors. Each edge in the graph represents one channel. The number in the edge annotation represents the capacity n of the channel [6]

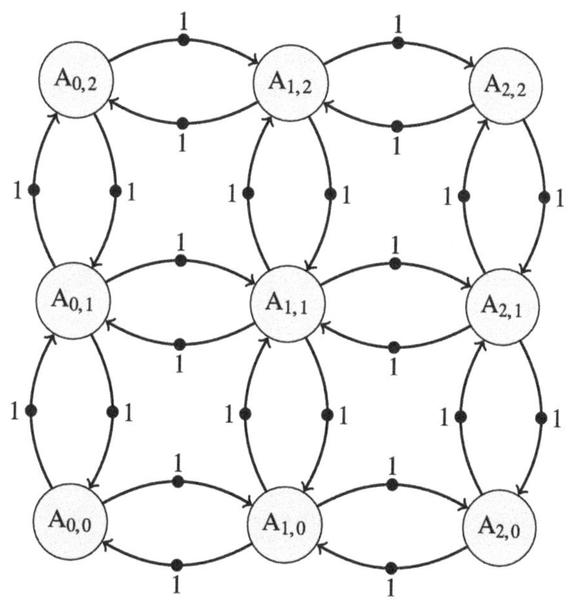

As an example, an SDF graph representations of a simulation run with 9 patches (or 3 × 3 actors) may be seen in Fig. 6.7. Further descriptions of the software and a performance evaluation are presented in [18].

6.3 Run-Time Application Mapping

As introduced already, hybrid application mapping combines a static DSE with a dynamic application mapping to guarantee user-defined requirements on different non-functional execution properties (e.g., latency or power consumption) of multiple applications being modeled each in ActorX10 and concurrently executed on a heterogeneous many-core architecture. In this section, we concentrate on the run-time mapping technique, which tries to find feasible mappings of the available operating points onto the target platform depending on the current system state and present an optimized problem-specific backtracking algorithm that helps efficiently solving this constraint satisfaction problem. Although HAM approaches significantly reduce the search space at run time, the decision of whether a feasible application mapping on a set of pre-occupied resources exists or not still is an NP-complete problem. However, experiments will show that the solving times for practically relevant problem sizes are on average a few milliseconds. Before explaining the details of the backtracking algorithm, some formal basic definitions used in HAM are summarized.

6.3.1 Mapping Based on Resource Constraints

In the following, we formally describe application mapping onto NoC-based many-core architectures and constraint graphs that represent the resource constraints of operating points.

Definition 6.7 ([7]) An *application graph* is a directed, bipartite graph $G_{\text{app}}(V, E)$ with vertices $V = T \cup M$ representing the set of application tasks T and the set of messages M. Each message $m \in M$ represents the data that needs to be exchanged between two communicating tasks. Directed edges $E \subseteq (T \times M) \cup (M \times T)$ model data dependencies. Edges exist either from a task $t \in T$ issuing a message $m \in M$ or from a message m to the task t receiving it. Figure 6.8a illustrates an example.

A heterogeneous NoC-based many-core architecture consists of a set of *processing elements* (PEs), possibly of different types, as well as routers, links, and *network interfaces* (NIs) that connect the PEs with the network routers. In the following, we consider the class of many-core systems with a mesh-based NoC interconnect topology as a representative example class of target architectures.

Definition 6.8 ([7]) A NoC architecture of width $\mathcal{X}$ and height $\mathcal{Y}$ is represented by an *architecture graph* $G_{\text{NoC}}(P, L)$ in which the set of vertices P model the PEs with $P = \{pe_{x,y} \mid 0 \leq x < \mathcal{X}, 0 \leq y < \mathcal{Y}\}$. Each PE is an instance from a set of available resource types $RES = \{r\_type_0, r\_type_1, \ldots, r\_type_{n-1}\}$. Function $type(pe_{x,y})$ gives the respective resource type of a PE $pe_{x,y} \in P$. Finally, the edges L model the NoC links between adjacent PEs.

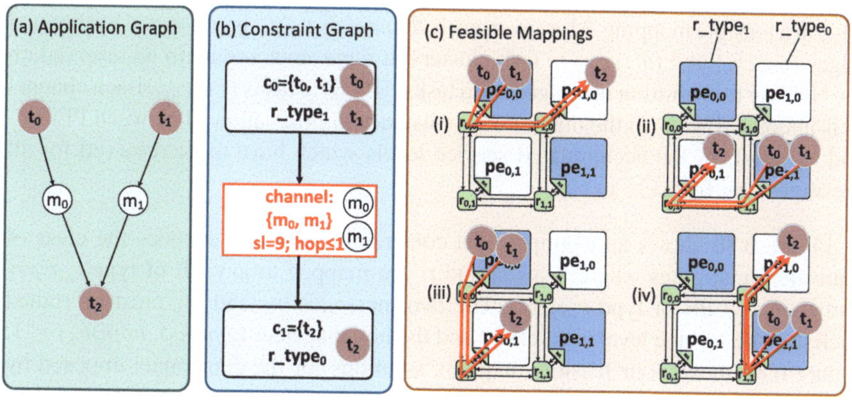

Fig. 6.8 Example of **a** an application graph, **b** an operating point with its corresponding constraint graph representing a clustering of tasks to resource types and constraints of the inter-task communication channels, and **c** four concrete mappings (embeddings) on a given architecture. Note that a constraint graph typically represents multiple feasible and in the objectives equivalent mappings [7]

In order to allow for a design-time timing analysis also for inter-task communications over the NoC, bandwidth reservation schemes are required. Although any reservation scheme that guarantees worst-case timing bounds is suitable here such as TDMA scheduling, we assume periodically scheduled traffic on each physical link using WRR [20] reservations. By assigning each message m routed over a set of physical links a certain number of time slots (denoted as *service level* in the following), best and worst-case end-to-end latencies may be guaranteed. The maximal number of service levels that can be assigned per link is denoted sl_{max} in the following.

The basic problem of mapping a given application graph G_{app} onto a given target architecture modeled by an architecture graph G_{NoC} consists in determining a binding (assignment) $\beta : T \rightarrow P$ of each task $t \in T$ to a PE $pe_{x,y} \in P$. Moreover, for each message $m \in M$, a corresponding route must be determined between the PEs where the sender and receiver task have been mapped. Messages between tasks which are bound onto the same PE can be exchanged locally. Messages between tasks on different PEs have to be routed over NoC resources. For routings, we assume XY-routing in the following w.l.o.g.

In the following, rather than trying to find static binding of tasks and routing of messages on the architecture, we follow a HAM approach as illustrated in Fig. 6.3: At design time, equivalence classes of feasible and Pareto-optimal mappings are explored each characterized by one operating point. First introduced in [4], an operating point carries a much lighter set of constraints with it to the run-time manager in the form of a constraint graph that is formalized as follows.

Definition 6.9 ([7]) A *constraint graph* is a bipartite graph $G_C(V_C, E_C)$ with a set of vertices $V_C = T_C \cup M_C$. Each vertex $c \in T_C$ represents a *task cluster* $c = \{t_0, \ldots, t_f\} \subseteq T$ of tasks that should be mapped onto the same PE, i.e., $\beta(t) = \beta(t'),\ \forall t, t' \in c$. Each $c \in T_C$ is annotated with a resource type r_type_i the PE must have in a feasible mapping. Moreover, each *channel* $\tilde{m} \in M_C$ is connected by directed edges $(c_{src}, \tilde{m})$ and $(\tilde{m}, c_{trgt})$ to task clusters. It represents a route to be reserved on the NoC for transmitting messages from tasks $t \in c_{src}$ to tasks $t' \in c_{trgt}$. Each channel $\tilde{m}$ is annotated with (a) the maximal hop distance $hop(\tilde{m})$ allowed between PEs c_{src} and c_{trgt} and (b) the accumulated service levels which have to be reserved for all messages sent from c_{src} to c_{trgt}.

Figure 6.8b shows an example of a constraint graph that encodes the class of equivalent mappings where tasks t_0 and t_1 are mapped to any PE of type r_type_1 and t_2 to any PE of type r_type_0. The two messages m_0 and m_1 must be routed such that the service level is at least 9 and the hop distances $hop(m_0), hop(m_1) \leq 1$. Figure 6.8c shows four feasible mapping solutions for the constraints imposed by the constraint graph.

6.3.2 *Backtracking Mapping Algorithm*

The algorithm presented in this section tries to find a feasible mapping β of a given constraint graph G_c on either an empty or a pre-occupied architecture graph G_{NoC} by applying an optimized backtracking technique, which exploits problem-specific knowledge (see Algorithm 6.1). It is designed to allow the parallel solving of up to k parallel instances by one thread each. Each instance is identified by the parameter *id* and operates on a copy of the architecture graph G_{NoC}.

The algorithm simultaneously operates on different task cluster orders $T_{c,i}$, where i indicates the currently selected order. If the backtracking search counted n_{max} unsuccessful tries to find a feasible mapping, the task cluster order is switched in a round-robin manner (line 42). This technique prevents the algorithm from stalling in an unfavorable task cluster order. In particular, a task cluster order by hop distance as well as all possible topological orders of the task clusters are used and initialized in lines 2 to 4. Furthermore, the algorithm uses data structures that are private to each task cluster order and subscripted by the index i. These are an architecture data structure pe_i, the mapping β_i, the current task cluster index j_i as well as task cluster specific data such as the current $x_{i,j}$ and $y_{i,j}$ coordinate, which are initialized in lines 9 to 13. $T_{c,i}(j)$ is used to return the j-th element of the task cluster order $T_{c,i}$ and $pe_i(x_{i,j}, y_{i,j})$ for getting the resource addressed by the given coordinate.

The backtracking algorithm works as follows: The current task cluster $T_{c,i}(j)$ is fetched according to the active task cluster order $T_{c,i}$ and the current resource $pe_i(x_{i,j}, y_{i,j})$ is selected according to a linear scanning order of the architecture's rectangular coordinate space (line 18). First, the availability of the resource is checked (line 19). Furthermore, the task and communication constraints for this task cluster—resource pair are checked (lines 20 to 23). This includes a type check and a check, whether the hop distance as well as the service level constraints to already mapped connected task clusters are fulfilled. The region check is part of an optimization and will be explained later. If a constraint could not be fulfilled, the next resource

Algorithm 6.1 Backtracking mapper algorithm.

1 **function** BACKTRACKINGEMBED(G_c,G_{NoC}, id, k)
2 $T_{c,0} = sortByHopDistance(T_c)$
3 **for** $i \leftarrow 0, |T_c| - 1$ **do**
4 $T_{c,i+1} = sortTopologically(T_c, i)$
5 $p = partition(\mathcal{X}, \mathcal{Y}, k)$
6 **for** $i \leftarrow 0, |T_c|$ **do**
7 **for** $j \leftarrow 0, |T_c| - 1$ **do**
8 **if** $j = 0$ **then**
9 $x_{i,j} = 0;\ x_{i,j}^{min} = p(id)_{x_{min}};\ x_{i,j}^{max} = p(id)_{x_{max}}$
10 $y_{i,j} = 0;\ y_{i,j}^{min} = p(id)_{y_{min}};\ y_{i,j}^{max} = p(id)_{y_{max}}$
11 **else**
12 $x_{i,j} = 0;\ x_{i,j}^{min} = 0;\ x_{i,j}^{max} = \mathcal{X} - 1$
13 $y_{i,j} = 0;\ y_{i,j}^{min} = 0;\ y_{i,j}^{max} = \mathcal{Y} - 1$
14 $i = j = n = 0$
15 **while** *true* **do**
16 **for** $y_{i,j} \leq y_{i,j}^{max};\ y_{i,j} = y_{i,j} + 1$ **do**
17 **for** $x_{i,j} \leq x_{i,j}^{max};\ x_{i,j} = x_{i,j} + 1$ **do**
18 $c = T_{c,i}(j);\ pe = pe_i(x_{i,j}, y_{i,j})$
19 **if** $\exists e : (e, pe) \in \beta_i$ **then continue** 17
20 **if** $type(pe) \neq type(c)$ **then continue** 17
21 **if** $\neg checkHopDistances(c, pe)$ **then continue** 17
22 **if** $\neg checkRegions(c, pe)$ **then continue** 17
23 **if** $\neg checkServiceLevel(c, pe)$ **then continue** 17
24 $\beta_i = \beta_i \cup \{(c, pe)\}$
25 $mapServiceLevel(c, pe)$
26 $updateRegions(c, pe)$
27 **if** $j = |T_c| - 1$ **then return** true
28 $j = j + 1;\ x_{i,j} = x_{i,j}^{min};\ y_{i,j} = y_{i,j}^{min}$
29 **continue** 15
30 $x_{i,j} = x_{i,j}^{min}$
31 $y_{i,j} = y_{i,j}^{min}$
32 **if** $j = 0$ **then return** false
33 **if** $n < n_{max}$ **then**
34 $j = j - 1;\ n = n + 1$
35 $c = T_{c,i}(j);\ pe = pe_i(x_{i,j}, y_{i,j})$
36 $\beta_i = \beta_i \setminus \{(c, pe)\}$
37 $unmapServiceLevel(c, pe)$
38 $restoreRegions(c, pe)$
39 $x_{i,j} = x_{i,j} + 1$
40 **else**
41 $n = 0$
42 $j_i = j;\ i = (i + 1) \bmod (|T_c| + 1);\ j = j_i$

is selected. If all checks have been evaluated positively, the task cluster is mapped on this resource (lines 24 to 26). This includes adding an edge to the mapping β_i and reserving time slots on all links along the paths between the current resource and the resources of already mapped connected task clusters. Updating the regions will be explained later. If all task clusters could be mapped successfully (line 27), a feasible mapping has been found and the algorithm terminates. Otherwise, the mapping process continues with the next task cluster. If no feasible mapping could be found for task cluster $T_{c,i}(j)$, the previous task cluster $T_{c,i}(j-1)$ is unmapped from its current resource (lines 36 to 38) and the mapping process continues with this task cluster at its next resource. If the first task cluster could not be mapped (line 32), no feasible mapping exists and the algorithm terminates.

One technique to exploit problem-specific knowledge is to restrict the possible task cluster resource candidates according to the hop distance constraints of already mapped connected task clusters, instead of scanning all available resources. A naïve implementation would introduce a lot of overhead due to managing lists of resources. Hence, we introduce rectangular *regions* of possible resource candidates, which are subsets of the architecture's coordinate space $\mathcal{X} \times \mathcal{Y}$. These regions are specific to each task cluster and have to be adapted dynamically. They are defined by the lower and upper bounds $x_{i,j}^{min}$, $y_{i,j}^{min}$, $x_{i,j}^{max}$, and $y_{i,j}^{max}$ and are initialized with the full coordinate space (lines 12 to 13). In case of parallel execution, the full coordinate space is partitioned into k distinct rectangular partitions (line 5) and each thread initializes its first task cluster region accordingly (lines 9 to 10). The regions are then used to scan only those resources, which are located within its bounds (lines 16 to 17). If a task cluster has been successfully mapped, the regions of not yet mapped connected task clusters have to be updated (line 26). This includes pushing the current bounds on a task cluster specific stack and restricting them according to the given hop distance constraint. If a task cluster is going to be unmapped, the previous regions of not yet mapped connected task clusters have to be restored by popping them from their stacks (line 38). Finally, the mapping of a task cluster on a resource can be excluded, if the potential new regions of not yet mapped connected task clusters and their current regions are disjoint (line 22).

6.3.3 Experimental Evaluation

In this section, we analyze the execution time needed for finding a feasible mapping using the introduced mapping algorithm (Algorithm 6.1). We implemented it in X10 and determined the execution time through simulation on different many-core architectures using InvadeSIM. We took test applications from the *embedded system synthesis benchmarks suite* (E3S) [21]. These applications stem from the domains of Automotive/Industrial (18 tasks, 21 messages), Telecom (14 tasks, 20 messages) and Consumer (11 tasks, 12 messages). They have been analyzed by the DSE, which optimizes for the objectives of application latency and the number of allocated resources and delivers a set of 100 operating points for each application.

We consider three different architectures varying in the NoC size (8×8, 12×12) and PE allocation (random, symmetric). Here, *symmetric* means that the different PE types are arranged in a symmetrical pattern while *random* means a random distribution. Each NoC is composed of three different resource types from the E3S benchmark suite, which vary in frequency. InvadeSIM allows modeling such architectures flexibly in terms of different NoC and PE configurations including the *clock frequency* of the processors and the NoC routers as well as the maximum amount of service levels available on each NoC link.

6.3.3.1 Full Resource Availability

First, we evaluated the execution (solving) times of the backtracking algorithm when no other application pre-occupies resources on the architecture. Constraint solving was performed for all 100 operating points of each application (*auto*, *telecom*, *consumer*) onto all three NoC architectures (8×8_*sym*, 8×8_*rand*, 12×12_*sym*). Parallelism was exploited by simultaneously running $th = 1, 4, 9$, and 16 instances of the solver on different places, where the first task cluster of each solver was initialized with another partition of the into k distinct partitions divided coordinate space. The frequencies of the processors of the simulated architecture were configured to 1, 000 MHz.

Figure 6.9 depicts the average execution times of the algorithm in ms. The overall execution time consists of the time required to initialize the parallel solvers by transferring the constraint graph (*init time*, shown in purple color) and the time for solving the constraints to obtain a feasible mapping (*solving time*, shown in blue color). It can be seen that (1) the overall execution time grows with the NoC size and (2) the dominant part of the solver is the init time: The init time grows with the degree of parallelism, because more data has to be transmitted to other places. However, it is still less than 3 ms in average for all test scenarios. This amount of timing overhead is affordable when admitting a new application to a many-core architecture. Figure 6.10 just shows the average solving time per experiment. It can be seen that the execution time can be reduced by parallelization.

6.3.3.2 Pre-occupied Resources

The previous section only considered empty architectures where no resources were pre-occupied. In this section, we study, how already occupied resources by other running applications may influence the execution time required to find a feasible mapping, if it exists. Typically, the success rate decreases in such cases. We vary the amount of pre-occupied resources from 10 to 100% and again evaluate the execution time of the backtracking algorithm for all test cases to find a mapping. Figure 6.11a and b summarize the execution times measured for the constraint solver for the three applications on the 8×8 rand architecture under a varying amount of pre-occupied resources.

Figure 6.11a shows the median and the 97.5% quantile, i.e., the worst-case execution time for solving 50 and 97.5% of all constraint graphs. It can be seen that there is an almost constant execution time until a resource utilization of around 40%. From then on, the success rate drops and less feasible mappings exist. However, the

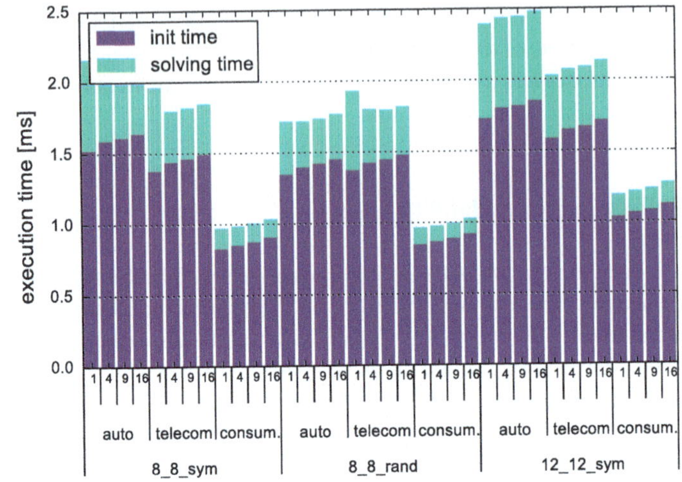

Fig. 6.9 Average execution times of the backtracking algorithm (including initialization and solving) for three benchmarks and differently sized MPSoC architectures [7]

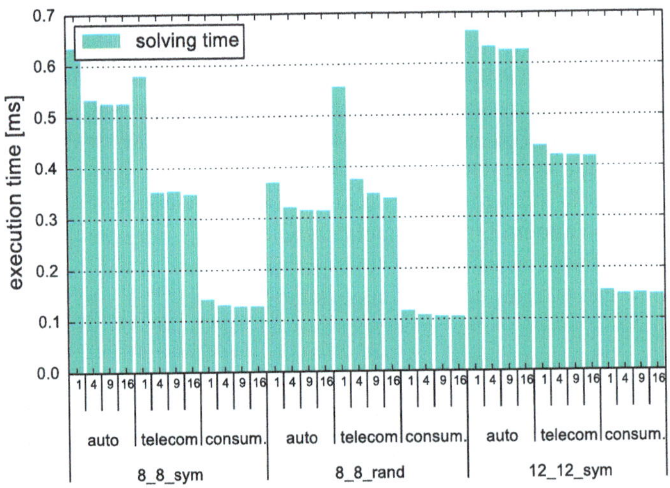

Fig. 6.10 Average execution times of the backtracking algorithm (only solving time) for three benchmarks and differently sized MPSoC architectures [7]

solver has to search the complete search space to verify no feasible mapping exists, which requires more time. Once the architecture is occupied for around 80–90%, the execution time is drastically reduced, because of a so-called precheck. In this check, the number of available resources is compared with the requirements from the constraint graph. Infeasibility can be easily deduced, if not enough resources are available and the backtracking algorithm even does not have to be started. Overall, constraint solving takes less than 1 ms in 97.5% of the scenarios.

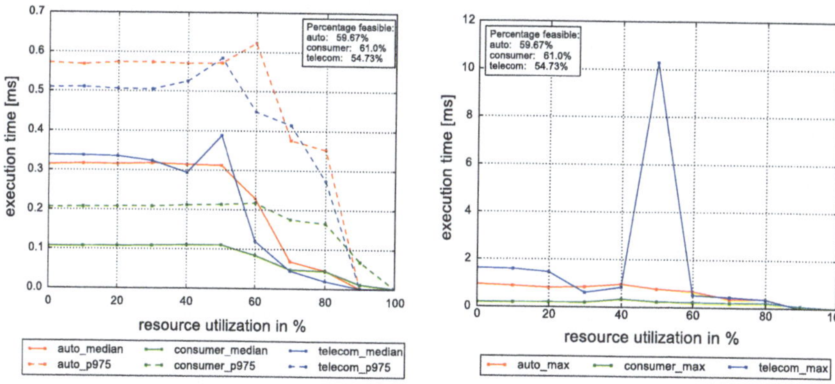

(a) Median and 97.5% quantile of the execution times

(b) Worst-case execution times

Fig. 6.11 Execution times observed under varying resource utilizations of the 8 × 8 rand architecture

Figure 6.11b shows the observed worst-case execution times. Here, a maximum of around 11 ms was observed on the 8 × 8 NoC. Response times of more than 100 ms are not tolerable for a run-time resource manager and a *timeout mechanism* has to be applied. The algorithm is stopped after the expiration of the timeout interval and the currently tested mapping is classified as infeasible. The timeout value has to be selected carefully, If it is chosen too low, the number of feasible mappings which are classified as infeasible may be increased. However, only 2.5% of our experiments would have been affected by a timeout value of 10 ms.

6.4 Simulative Predictability Analysis

In this section, we present two simulation-based case studies to analyze and compare the ability of invasive computing to guarantee given user requirements and to reduce execution time and throughput jitter of soft real-time stream processing applications to a minimum on heterogeneous NoC-based MPSoCs by isolating applications from each other. The first case study considers isolation by exclusive reservation of computing and I/O tiles for each application and the second by guaranteeing a minimal NoC bandwidth between communicating tasks of an application.

As an example of a complex streaming application with throughput requirements, we analyzed the actor graph of an object detection algorithm chain as introduced in Sect. 6.2.3.1. Since object detection algorithms are often used to detect objects in a video stream in real time (e.g., in robotics to control the movement of a robot), a minimum throughput often has to be guaranteed to fulfill a certain task. A brief overview of the individual tasks in an object detection algorithm chain in SDF graph semantics is depicted in Fig. 6.12. Here, a task is activated as soon as a token is on

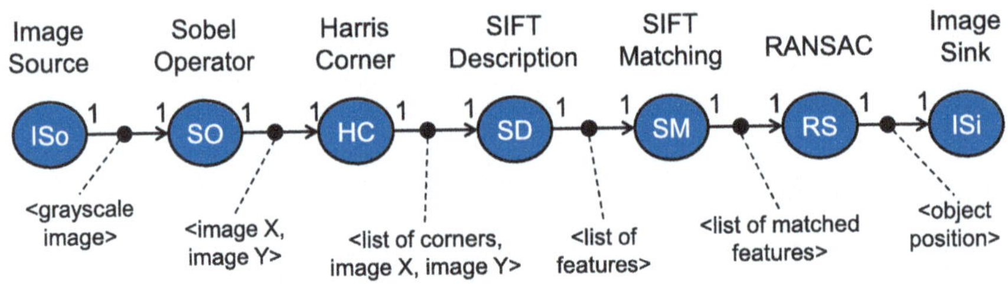

Fig. 6.12 Actor model with synchronous data flow graph semantics of the object detection algorithm chain

its input edge and puts a token on its output edge afterwards. The *image source* task has no input edge, it generates tokens with a given rate.

The object detection algorithm provides a lot of parallelism in case of processing a stream of images, e.g., delivered by a camera. Pipeline parallelism can be exploited, which allows all of the tasks running in parallel while each is processing a different image instance and subsequently sending data to the next stage and waiting for new data from the previous stage. This object detection chain was implemented in *ActorX10*, encapsulating each of the object detection tasks by an actor. Each actor communicates via FIFO channels with its neighbors. ActorX10 takes care of the communication of tokens (messages) from one actor to another by allocating the buffer in shared memory if both actors are mapped to the same tile or transmitting it over the NoC if the actors reside on different tiles.

The execution time jitter is determined by the difference between best-case and worst-case execution time observed in all simulation scenarios and is a predictability marker (see Sect. 6.1.1) of the objective latency L, which denotes the time it takes to process one image frame by the full pipeline depicted in Fig. 6.12. The throughput Tr is measured in FPS and denotes the frequency of how many frames can be processed by the image processing pipeline per second.

6.4.1 *Exclusive Tile Reservation*

In order to demonstrate the advantages of invasive computing in terms of predictable execution times in dynamic workload scenarios, in our first case study, we compare a regular execution of a mix of streaming applications on a heterogeneous MPSoC to executions with previously analyzed resource constraints and exclusive reservations of tiles through invasion. We show that the introduced HAM approach is able to guarantee given user requirements for streaming applications such as throughput (FPS) and to increase predictability in terms of jitter in latency and throughput, whereas a regular execution does not only violate user requirements, but also degrades predictability.

The algorithm chain from Fig. 6.12 was analyzed regarding throughput for a parallel execution on a heterogeneous MPSoC. The architecture contains five quad core RISC tiles, two I/O tiles for input and output, one global memory tile, and one TCPA accelerator tile, such as depicted in Fig. 6.13. The application needs two I/O tiles and four RISC tiles as communication and computation resources to ensure user requirements, i.e., an average throughput of 25 FPS. At run time, resources have to be reserved according to the constraints for executing the application. The user requirement and a predictable execution in the scope of algorithmic variability is guaranteed, if the reservation (invasion) is successful, even if other applications are running on the architecture.

In our first scenario according to Fig. 6.13a, we consider a regular (non-invasive) execution of the object detection application together with a motion planning application and an FIR filter. Twelve images are filled into the actor chain and are simulated to obtain latency (to process one image) and throughput values as well as different variations of these values (jitter). The performance results are determined by *InvadeSIM*, which allows modeling such heterogeneous architectures and simulating parallel applications on them. Due to non-exclusive execution, the other applications may be mapped on the same tiles where tasks of the object detection application are running, as shown in Fig. 6.13a. In this case, the execution of the object detection application is disturbed by the other tasks. Our second scenario (see Fig. 6.13b) considers an invasive execution, where each application is executed isolated from each other after invasion. The results of the simulation of both scenarios are shown in the following table. It can be seen that for the non-invasive case, the user requirement of 25 FPS cannot be fulfilled. Furthermore, the jitter for throughput is maximum 30% and for latency 29%. In case of an invasive execution, the user requirement of 25 FPS is achieved, the jitter for throughput is maximum 18% and for latency 7%. These results show that invasive computing helps to meet user requirements and to increase predictability in terms of variations of latency and throughput of complex stream processing applications.

	Non-invasive execution	Invasive execution
Average latency	180 ms	92 ms
Latency jitter	[128, 205 ms]	[86, 100 ms]
Average throughput	17.9 FPS	25 FPS
Throughput jitter	[13.7 FPS, 23.3 FPS]	[20.4 FPS, 29.3 FPS]

6.4.2 NoC-Bandwidth Reservation

In order to show the advantages of invasive computing in terms of reducing execution time and throughput jitter by guaranteeing a minimal NoC bandwidth between communicating tasks of an application, in our second case study, we consider an

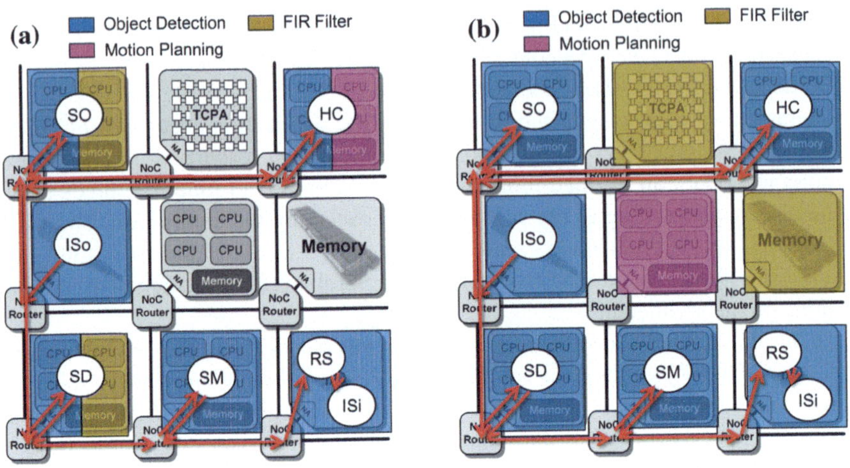

Fig. 6.13 Non-invasive execution of a mix of streaming applications on the left side: The FIR filter is overlapping with the object detection task. Execution with exclusive resource reservation on the right side: All task are executed isolated from each other [8]

application scenario, where the object detection application is executed concurrently with several communicating and parallel Monte Carlo simulations calculating an approximation of the number π. The target architecture is a 4×4 NoC-based architecture comprising of different types of tiles as well as different types of processors on these tiles. One mapping of the applications on the target architecture is shown in Fig. 6.14. Additionally, the used NoC links between the communicating tasks of each application are highlighted in color. It can be seen that there are overlapping paths between the object detection application and the Monte Carlo simulations (all horizontal links in the second column of the tile-based architecture). Here, we try to minimize the execution time and throughput jitter of the object detection task chain by applying invasive computing principles to guarantee a minimal bandwidth between its communicating tasks.

As introduced in Chap. 4, the invasion of NoC resources [22] is realized through state-of-the-art *virtual channels* (VCs) and a *weighted round-robin* (WRR) arbitration [20]. An application can reserve a certain NoC budget $SL(m)$ (also called *service level* (SL)) for a message m from the maximal available NoC budget of a link SL_{max}. The budget refers to *time slots* (TSs) which are assigned to messages in the interval of SL_{max}. During invasion, the availability of TSs and VCs is checked and then assigned to the communicating tasks according to their request. If a message does not utilize its reserved time slot, it can be used by another message.

In the following experiments, we simulated the described application scenario on the 4×4 NoC architecture from Fig. 6.14 using *InvadeSIM*, which is also able to simulate certain NoC bandwidth reservations on tile-based heterogeneous MPSoC. In case of a bandwidth requirement between the tasks of the object detection application, *guaranteed service* (GS) NoC channels with different service levels must be reserved

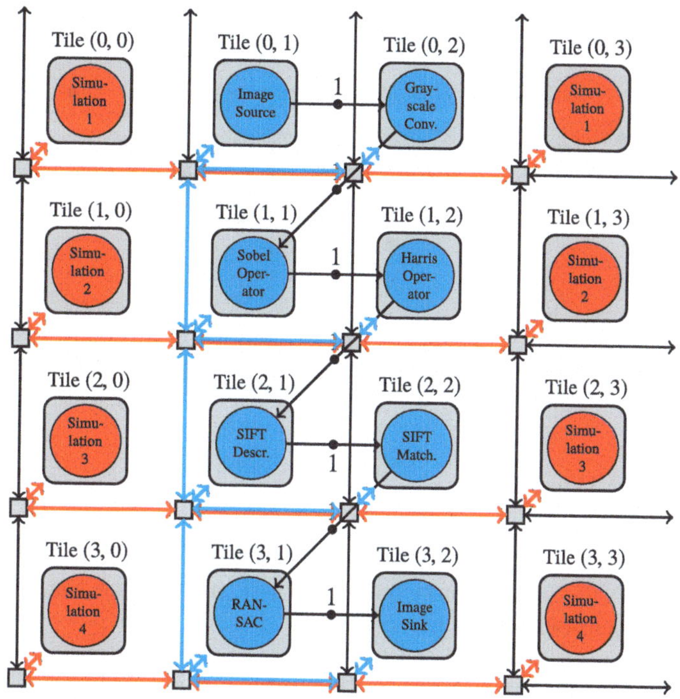

Fig. 6.14 Example of an application mapping of the object detection (blue) task chain as well as multiple Monte Carlo simulation instances (red) on a 4 × 4 NoC-based architecture [1]

for messages between communicating tasks, after the mapping of the tasks to tiles is done. In contrast, the Monte Carlo simulations only use *best effort* (BE) channels between communicating tasks. They are configured to claim up to eight BE transmissions in parallel to increase the utilization of the NoC. The observed latency and throughput variations can be seen in Figs. 6.15a and b. The service level reserved for each GS channel of the object detection application is plotted for values of $SL(m) = 1$ to $SL(m) = 8, \forall m$. In both cases, the objective values and predictability markers significantly improve when increasing the reserved communication bandwidth of messages: For the latency marker, $L(p, \{SL(m) = 1, \forall m\}, I) = [111.34, \ldots, 162.34]$ drops to $L(p, \{SL(m) = 6, \forall m\}, I) = [89.66, \ldots, 90.66]$, and for the throughput predictability marker from $Tr(p, \{SL(m) = 1, \forall m\}, I) = [18.88, \ldots, 20.65]$ to $Tr(p, \{SL(m) = 6, \forall m\}, I) = [24.02, \ldots, 25.63]$. If the user would formulate a requirement of an end-to-end latency of at most 100 ms, the operating point with $SL(m) = 5 \forall m$ could be preferred as further increasing the service levels does not further improve the throughput and its variation, but would reduce the bandwidth available for other applications. Still, the simulation-based evaluation only allows determining observed predictability markers and is therefore only applicable for soft user requirements.

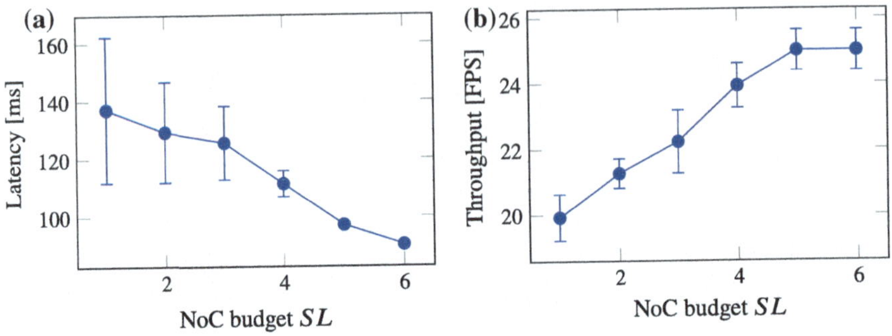

Fig. 6.15 Latency L and throughput Tr and their min/max values of the object detection task chain in dependence on different service levels SL allocated for inter-task communications [1]

6.5 Related Work

In the following, existing *actor libraries* comparable to ActorX10 are discussed. In contrast to X10, Scala innately includes a basic actor model and furthermore employs the open source actor library Akka [23], which offers powerful concepts such as hierarchical supervised and typed actors as well as actor migration upon failure, and to some extent also actor reallocation at run time. In StreamX10 [24], a streaming language called COStream is proposed to describe actor functionality and dependencies between the actors. This language is being compiled down to parallel X10 code by an external compiler. The mapping of actors to places is statically fixed and cannot be changed at run time. StreamIt [25] is also language/compiler combination for characterizing streaming applications, but aims to generate efficient code by applying stream-specific optimizations. Phaser Beams [26] is based on the Phasers construct of the Habanero-Java language, which is in turn derived from X10 clocks. However, this extension result in code that is not native X10 compatible. Actor functionalities are encapsulated in X10 method calls and their dependencies are specified by input and output accumulators, which are fixed at compile time. In comparison, our approach is native X10 compatible and does not require any extension or an additional compiler. It also supports dynamic migration of actors and enables analyses through the usage of state machines. In [27], we described the implementation of a task library that served as precursor to ActorX10. While this library already supports the notion of actors that are connected by channels, there is no notion of ports and no formalism is provided as well as no support for migration of actors at run time. Shali [28] proposed an actor-oriented extension of the programming language Chapel. Their approach also enables the developer to distribute actors onto different shared-memory domains. However, instead of ports and channels, mailboxes are used and they do not consider a dynamic migration of actors between places. Imam and Sarkar [29] presented a unified concept of the actor model and the task-parallel async-finish model in Habanero-Java as well as Habanero-Scala and demonstrated the benefits of this combined model.

Run-time mapping approaches are particularly beneficial in scenarios of dynamically changing workloads since they determine mappings of applications onto MPSoCs at run-time. However, it may become very expensive in terms of solving time to determine exact solutions. Therefore, fast heuristics are typically used such as simple nearest neighbor algorithms as proposed in [30, 31]. Such approaches often ignore or cannot guarantee non-functional execution qualities on the resulting application mappings. Typical objectives considered in run-time optimization are soft real-time as presented in [32], energy as shown in [33], or average speedup as presented in [34]. In [35], a two-step iterative application mapping technique for heterogeneous MPSoCs is proposed. After an initial greedy task to resource assignment has been applied, an optimization step tries to improve the mapping. If the QoS requirements are not met, the assignment is marked as infeasible and another initial greedy mapping is selected. In contrast, our run-time mapping algorithm works with pre-defined mapping classes (constraint graphs), which have been determined at design time and guarantee the QoS requirements once embedded.

6.6 Summary

Recently, HAM approaches have obtained big attention as a promising methodology for application mapping on many-core systems. The major idea and benefit of HAM is to perform time-intensive parts of complex optimization and verification including DSE of Pareto-optimal mapping classes at design time and leave a smaller problem to determine a concrete mapping of application tasks and communications based on a set of mapping constraints to the run-time resource manager. In this chapter, two major contributions to HAM were proposed: An implementation of the actor model in X10 as a library called ActorX10. By making the interactions between concurrently executed strands of code explicit, the library simplifies reasoning about the concurrent behavior and data flow in application codes significantly. Two concrete real-world examples from the multimedia and the high-performance computing domain have been successfully modeled using ActorX10. These examples serve to demonstrate the expressiveness and generality of our approach as well as the synergy effects that may be gained by fusing the APGAS paradigm with an actor-based programming model. As second major contribution, we proposed an exact approach to solve the CSP by a problem-specific backtracking solver. It was shown that the backtracking algorithm is able to determine feasible mappings at run time within a few milliseconds also for many-core systems with more than 100 processors. Furthermore, we could demonstrate the capabilities of invasive computing to provide predictable execution of stream processing applications on heterogeneous MPSoCs and to guarantee statically analyzed best and worst-case timing requirements on latency and throughput.

References

1. J. Teich et al., Language and compilation of parallel programs for *-Predictable MPSoC execution using invasive computing, in *Proceedings of the 10th IEEE International Symposium on Embedded Multicore/Manycore Systems-on-Chip (MCSoC)*, Lyon, France, 21–23 September 2016. IEEE, pp. 313–320. ISBN: 978-1-5090-3531-1. https://doi.org/10.1109/MCSoC.2016.30
2. A. Weichslgartner et al., *Invasive Computing For Mapping Parallel Programs to Many-Core Architectures*. Computer Architecture and Design Methodologies, 15 January 2018 (Springer, Singapore, 2018). ISBN: 978-981-10-7355-7. https://doi.org/10.1007/978-981-10-7356-4
3. T. Schwarzer et al., Symmetry-eliminating design space exploration for hybrid application mapping on many-core architectures. IEEE Trans. Comput.-Aided Design Integr. Circuits Syst. (TCAD) **37**(2), 297–310 (2018). ISBN: 0278-0070. https://doi.org/10.1109/TCAD.2017.2695894
4. A. Weichslgartner et al., DAARM: design-time application analysis and run-time mapping for predictable execution in many-core systems, in *Proceedings of the International Conference on Hardware/Software Codesign and System Synthesis (CODES+ISSS)*, 12–17 October 2014 (New Delhi, India), pp. 34:1–34:10. ACM. ISBN: 978-1-4503-3051-0. https://doi.org/10.1145/2656075.2656083
5. B. Akesson et al., Composability and predictability for independent application development, verification, and execution, in *Multiprocessor System-on-Chip: Hardware Design and Tool Integration*, ed. by M. Hübner, J. Becker (Springer, New York, 2011), pp. 25–56. ISBN: 978-1-4419-6460-1. https://doi.org/10.1007/978-1-4419-6460-1_2
6. S. Roloff et al., ActorX10: an actor library for X10, in *Proceedings of the 6th ACM SIGPLAN Workshop on X10 (X10)*, Santa Barbara, CA, USA, 14 June 2016, pp. 24–29. ISBN: 978-1-4503-4386-2. https://doi.org/10.1145/2931028.2931033
7. T. Schwarzer et al., On the complexity of mapping feasibility in many-core architectures, in *Proceedings of the 12th IEEE International Symposium on Embedded Multicore/Many-Core Systems-on-Chip (MCSoC)*, Hanoi, Vietnam, 12–14 September 2018. IEEE
8. S. Roloff et al., Invasive computing for predictable stream processing: a simulation-based case study, in *Proceedings of the 13th IEEE Symposium on Embedded Systems for Real-time Multimedia (ESTIMedia)*, Amsterdam, The Netherlands, 8–9 October 2015. IEEE. ISBN: 978-1-4673-8164-2. https://doi.org/10.1109/ESTIMedia.2015.7351761
9. P. Axer et al., Building timing predictable embedded systems. ACM Trans. Embedded Comput. Syst. (TECS) **13**(4), 82:1–82:37 (2014). ISSN: 1539-9087. https://doi.org/10.1145/2560033
10. D. Grund, J. Reinecke, R. Wilhelm, A template for predictability definitions with supporting evidence, in *Bringing Theory to Practice: Predictability and Performance in Embedded Systems (PPES)*, Grenoble, France, ed. by P. Lucas et al. Open Access Series in Informatics (OASIcs), vol. 18. Schloss Dagstuhl-Leibniz-Zentrum fuer Informatik, 22–31 March 2011, pp. 22–31. ISBN: 978-3-939897-28-6. https://doi.org/10.4230/OASIcs.PPES.2011.22
11. L. Thiele, R. Wilhelm, Design for timing predictability. Real-Time Syst. (RTS) **28**(2), 157–177 (2004). ISSN: 1573-1383. https://doi.org/10.1023/B:TIME.0000045316.66276.6e
12. T. A. Henzinger, Two challenges in embedded systems design: predictability and robustness. Philos. Trans. R. Soc. Lond. A Math. Phys. Eng. Sci. **366**, 3727–3736 (1881). ISSN: 1364-503X. https://doi.org/10.1098/rsta.2008.0141
13. C. Hewitt, P. Bishop, R. Steiger, A universal modular ACTOR formalism for artificial intelligence, in *Proceedings of the 3rd International Joint Conference on Artificial Intelligence (IJCAI)*, 20–23 August 1973 (Morgan Kaufmann Publishers Inc., Stanford, 1973), pp. 235–245
14. G. A. Agha, *Actors: A Model of Concurrent Computation in Distributed Systems*. Technical report AITR-844. MIT Artificial Intelligence Laboratory, June 1985
15. J. Falk et al., Integrated modeling using finite state machines and dataflow graphs, in *Handbook of Signal Processing Systems*, ed. by S. S. Bhattacharyya et al. (Springer, Boston, 2010), pp. 1041–1075. ISBN: 978-1-4419-6345-1. https://doi.org/10.1007/978-1-4419-6345-1_36

16. K. Strehl et al., FunState—an internal design representation for codesign. IEEE Trans. Very Large Scale Integr. (VLSI) Syst. **9**(4), 524–544 (2001). ISSN: 1063-8210. https://doi.org/10.1109/92.931229
17. L. Thiele et al., FunState—an internal design representation for codesign, in *Proceedings of the IEEE/ACM International Conference on Computer-Aided Design (ICCAD)*, San Jose, CA, USA, 7–11 November 1999. IEEE, pp. 558–565. ISBN: 0-7803-5832-5. https://doi.org/10.1109/ICCAD.1999.810711
18. A. Pöppl, M. Bader, SWE-X10: an actor-based and locally coordinated solver for the shallow water equations, in *Proceedings of the 6th ACM SIGPLAN Workshop on X10 (X10)*, Santa Barbara, CA, USA, 14 June 2016. ACM, pp. 30–31. ISBN: 978-1-4503-4386-2. https://doi.org/10.1145/2931028.2931034
19. A. Breuer, M. Bader, Teaching parallel programming models on a shallow-water code, in *Proceedings of the 11th International Symposium on Parallel and Distributed Computing (ISPDC)*, Munich, Germany, 25–29 June 2012. IEEE, pp. 301–308. ISBN: 978-1-4673-2599-8. https://doi.org/10.1109/ISPDC.2012.48
20. J. Heisswolf, R. König, J. Becker, A scalable NoC router design providing QoS support using weighted round robin scheduling, in *IEEE 10th International Symposium on Parallel and Distributed Processing with Applications (ISPA)*, Leganes, Spain, 10–13 July 2012. IEEE, pp. 625–632. ISBN: 978-1-4673-1631-6. https://doi.org/10.1109/ISPA.2012.93
21. R. Dick, *Embedded System Synthesis Benchmarks Suite (E3S)* (2010). http://ziyang.ewecs.umich.edu/dickrp/e3s/
22. J. Heisswolf et al., The invasive network on chip-a multi-objective many-core communication infrastructure, in *Workshop Proceedings of the 27th International Conference on Architecture of Computing Systems (ARCS)*25–28 February 2014. (VDE Verlag, Lübeck, 2014), pp. 1–8. ISBN: 978-3-8007-3579-2
23. Lightbend Inc. *Akka–Scala Actor Library*, 8 May 2018. http://akka.io
24. H. Wei et al., StreamX10: a stream programming framework on X10, in *Proceedings of the 2nd ACM SIGPLAN X10 Workshop (X10)*, Beijing, China, 14 June 2012. ACM, pp. 1:1–1:6. ISBN: 978-1-4503-1491-6. https://doi.org/10.1145/2246056.2246057
25. W. Thies, M. Karczmarek, S. P. Amarasinghe, StreamIt: a language for streaming applications, in *Proceedings of the 11th International Conference on Compiler Construction (CC)*, Grenoble, France, ed. by R. Nigel Horspool, vol. 2304. Lecture Notes in Computer Science (LNCS) (Sringer, Heidelberg, 2012) 8–12 April 2002, pp. 179–196. ISBN: 978-3-540-43369-9. https://doi.org/10.1007/3-540-45937-5_14
26. J. Shirako et al., Phaser beams: integrating stream parallelism with task parallelism, in *Presentation at the 1st ACM SIGPLAN X10Workshop (X10)*, 4 June 2011
27. S. Roloff, F. Hannig, J. Teich, Towards actor-oriented programming on PGAS-based multicore architectures, in *Workshop Proceedings of the 27th International Conference on Architecture of Computing Systems (ARCS)*, 25–28 February 2014 (VDE Verlag, Lübeck, 2014), pp. 1–2. ISBN: 978-3-8007-3579-2
28. A. Shali, C. Lin, Actor oriented programming in chapel (2010)
29. S. M. Imam, V. Sarkar, Integrating task parallelism with actors, in *Proceedings of the ACM International Conference on Object Oriented Programming Systems Languages and Applications (OOPSLA)*, Tucson, AZ, USA, October 2012. ACM, pp. 753–772. ISBN: 978-1-4503-1561-6. https://doi.org/10.1145/2384616.2384671
30. E. Carvalho, N. Calazans, F. Moraes, Heuristics for dynamic task mapping in NoC-based heterogeneous MPSoCs, in *Proceedings of the 18th IEEE/IFIP International Workshop on Rapid System Prototyping (RSP)*, Porto Alegre, Brazil, 28–30 May 2007. IEEE, pp. 34–40. ISBN: 0-7695-2834-1. https://doi.org/10.1109/rsp.2007.26
31. A. Weichslgartner, S. Wildermann, J. Teich, Dynamic decentralized mapping of tree-structured applications on NoC architectures, in *Proceedings of the 5th IEEE/ACM International Symposium on Networks on Chip (NoCS)*, Pittsburgh, Pennsylvania, 1–4 May 2011. ACM, pp. 201–208. ISBN: 978-1-4503-0720-8. https://doi.org/10.1145/1999946.1999979

32. E. W. Brião, D. Barcelos, F. R. Wagner, Dynamic task allocation strategies in MPSoC for soft real-time applications, in *Proceedings of the Design, Automation and Test in Europe (DATE)*, Munich, Germany, 10–14 March 2008. IEEE, pp. 1386–1389. https://doi.org/10.1145/1403375.1403709
33. C.-L. Chou, U. Y. Ogras, R. Marculescu, Energy- and performance-aware incremental mapping for networks on chip with multiple voltage levels. IEEE Trans. Comput.-Aided Design Integr. Circ. Syst. (TCAD) **27**(10), 1866–1879 (2008). ISSN: 0278-0070. https://doi.org/10.1109/TCAD.2008.2003301
34. S. Kobbe et al., DistRM: distributed resource management for on chip many-core systems, in *Proceedings of the 9th IEEE/ACM/IFIP International Conference on Hardware/Software Co-design and System Synthesis (CODES+ISSS)*, Taipei, Taiwan, 9–14 October 2011. IEEE, pp. 119–128. ISBN: 978-1-4503-0715-4. https://doi.org/10.1145/2039370.2039392
35. P. K. F. Hölzenspies et al., Run-time spatial mapping of streaming applications to a heterogeneous multi-processor system-on-chip (MPSoC), in *Proceedings of the Design, Automation and Test in Europe (DATE)*, Munich, Germany, 10–14 March 2008. ACM, pp. 212–217. https://doi.org/10.1145/1403375.1403427

Chapter 7
Conclusions and Future Directions

In this chapter, we shortly summarize the goals and achievements of this book and possible future directions.

7.1 Conclusions

In this book novel simulation techniques have been investigated that enable the validation and variants' exploration of all essential features of a novel computing paradigm called invasive computing. Here, an application may claim a set of resources of a heterogeneous tiled architecture based on temporal computing demands and the availability of resources as well as their physical states such as temperature or power. Accordingly, the dynamic workload is then distributed and freed again after parallel execution to become available for other applications. A big challenge here is that the architectures under investigation are highly complex and diverse as well as the workloads to be simulated. To study the benefits of invasive computing, target architectures composed of dozens to hundreds of heterogeneous processor cores organized in computing tiles and interconnected via a NoC are to be simulated with multiple competing highly parallel applications as typical workload. To handle this complexity, in this book a novel high-level processor simulation approach, a hybrid NoC simulation approach as well as parallel simulation methods have been proposed and finally implemented into the simulation framework InvadeSIM. In order to verify performance goals of the heterogeneous MPSoC apart from functional correctness, we proposed a novel direct-execution simulation approach, which is based on a discrete-event simulation and a time-warping mechanism to scale the elapsed execution times on the simulation host to the simulated target. We proposed a performance counter measurement for uninterrupted code sections based on three events that are feed into a linear performance estimation model to determine a prediction of the target execution

S. Roloff et al., *Modeling and Simulation of Invasive Applications and Architectures*, Computer Architecture and Design Methodologies,
https://doi.org/10.1007/978-981-13-8387-8_7

time. Each processor type requires a calibration on the host machine. By evaluating several benchmark programs, we could demonstrate that our simulation approach results in a maximum slowdown of 5 for single-core simulation compared to native execution, while resulting in an error of 18%. Furthermore, a DES framework was proposed that easily allows integrating and synchronizing different simulation tasks such as software or hardware simulations. We showed the relation and the mapping of our synchronization approach to the runtime of a parallel programming language called X10. Different timing simulation models for hardware accelerators could be integrated easily as shown for the TCPA and the *i*-Core. A shared-memory bandwidth congestion model was proposed that allows simulating the slowdown effects in case of multiple concurrent accesses. Furthermore, we presented a technique to adapt the frequencies of the simulated cores dynamically. Besides the simulation of a single tile, a novel hybrid NoC simulation approach has been presented. This approach accelerates the simulation of NoC traffic. The main idea of this approach is to detect steady communication intervals during the transmission of data, which are then exploited to forward the simulation time. We introduced a so-called transmission graph, which allows determining the individual bandwidth values for each transmission and forwarding the NoC simulation time accordingly in intervals between arrivals and terminations of transmission requests coming from the simulated applications. Furthermore, we presented an integration technique to incorporate this NoC simulation approach into InvadeSIM. Here, we investigated the underlying communication mechanism of X10 to extract the network traffic from X10 programs to feed the NoC simulator. Experiments using distributed applications from the IMSuite benchmark showed an average simulated latency error of less than 5%, while accelerating the simulation time for the NoC simulation by a factor of 1,000 and many cases. In order to exploit the performance potential of modern multi-core host machines, we presented parallelization techniques for InvadeSIM. Proposed were four novel parallel discrete-event simulation techniques, which map activity-level parallelism within the applications to core-level parallelism on the target architecture and back to thread-level parallelism on the host machine. Experiments with parallel real-world applications have been used to compare the different techniques against each other. On a 12-core Intel Xeon processor, an average speedup of 8.2 compared to sequential simulation could be achieved for the entire set of evaluated parallel X10 applications. Besides the novel simulation techniques, the second main contribution of this book was to model invasive applications as actor graphs and use simulation for analyzing the predictable execution behavior and supporting HAMs. Two major contributions to HAM were proposed: First, the design and development of a novel actor-oriented programming library called ActorX10, which allows describing computation and communication of stream processing applications in X10 formally, which represents the fundamental application model used in HAM. Two concrete real-world examples from the multimedia and the high-performance computing domain have been successfully modeled using ActorX10. As second major contribution, we proposed an exact approach to solve the CSP by a problem-specific backtracking solver. It was shown that the backtracking algorithm is able to determine feasible mappings at run time within a few milliseconds for many-core systems with more than 100 processors.

Furthermore, we could demonstrate the capabilities of invasive computing to provide predictable execution of stream processing applications on heterogeneous MPSoCs and to guarantee statically analyzed best and worst-case timing requirements on latency and throughput.

7.2 Future Directions

The full-system simulator InvadeSIM is a major and operative instrument for validating and exploring different programming concepts and architecture variants. It has been used as testbed for first implementations of programming concepts such as InvadeX10 and ActorX10 as well as for application mapping approaches. This important role may be continued by supporting novel ideas and techniques, particularly on enforcement of non-functional properties of program execution. InvadeSIM's ultra-fast parallel simulation including precise *i*-NoC simulation is a prerequisite for the assessment of non-functional properties such as timing predictability. In this regard, InvadeSIM may be extended to model and simulate other system-relevant non-functional properties of a given execution platform, i.e., temperature, power, and energy. The evaluation of these endogenous system properties is essential for studying enforcement techniques, e.g., staying within a certain power or temperature corridor or for the early evaluation of different techniques for thermal, power, and dark-silicon management. To accomplish this goal, popular models and tools for temperature and power estimation need to be integrated into InvadeSIM. Candidates include tools like HotSpot for thermal analysis based on RC-network models [1], but also more efficient approaches such as MatEx [2], or eTPMon [3]. Similarly, in the case of power, well-known estimation techniques (e.g., regression-based [4] or power models at instruction level [5]) may be incorporated into InvadeSIM. Alternatively, since InvadeSIM already extensively utilizes performance counters for timing simulation, these might also serve as a basis for real-time power estimation [6]. While InvadeSIM's architecture description language allows specifying heterogeneous tiled architectures in a very flexible manner (number and type of compute tiles, various interconnect parameters, number of cores per tile, and many other parameters), only one fixed clock rate per processor core can be specified so far. In order to enable the simulation of commonly used power management features such as dynamic voltage and frequency scaling (DVFS) and low-power CPU sleep states, InvadeSIM's architecture description language might be extended by means to specify tables storing voltage/frequency pairs and power state machines (attributed by, e.g., average power or frequency), respectively. Based on the extensions mentioned above, InvadeSIM may provide a quick solution to experimentally develop and evaluate techniques for *run-time requirement enforcement* (RRE), which are based on so-called enforcement automata. Their capabilities are not only to guide the execution of a program in given corridors of timing and power but also to minimize energy on invaded resources if these bounds allow. Furthermore, the system properties mentioned above, including power and temperature determined by simulation, have to be lifted to the X10-program level to serve as input to the enforcement automata that shall be generated based on the analysis of a set of requirements and a given description of the target platform. Thus, the InvadeSIM framework would have to be extended accordingly. InvadeSIM could be the testbed for strict and loose as well as centralized and distributed enforcement techniques.

References

1. W. Huang et al., HotSpot: a compact thermal modeling methodology for early-stage VLSI design, in *IEEE Transactions on Very Large Scale Integration (VLSI) Systems 14.5* (2006), pp. 501–513. https://doi.org/10.1109/TVLSI.2006.876103
2. S. Pagani et al., MatEx: efficient transient and peak temperature computation for compact thermal models, in *18th Design, Automation and Test in Europe (DATE), Grenoble, France, 9–13 Mar 2015*. (2015), pp. 1515–1520. https://doi.org/10.7873/DATE.2015.0328
3. E. Glocker et al., Emulation of an ASIC power and temperature monitoring system (eTPMon) for FPGA prototyping, in *Microprocessors and Microsystems 50* (2017), pp. 90–101. ISSN: 0141-9331. https://doi.org/10.1016/j.micpro.2017.03.001
4. A. Bogliolo, L. Benini, G. De Micheli, Regressionbased RTL power modeling. ACM Trans. Des. Autom. Electron. Syst. (TODAES) **5**(3), 337–372 (2000). https://doi.org/10.1145/348019.348081
5. S. Sultan, and S. Masud, Rapid software power estimation of embedded pipelined processor through instruction level power model, in *International Symposium on Performance Evaluation of Computer Telecommunication Systems (SPECTS), Istanbul, Turkey, 13–16 July 2009*, vol. 41 (IEEE, 2009), pp. 27–34
6. K. Singh, M. Bhadauria, S.A. McKee, Real time power estimation and thread scheduling via performance counters, in *ACM SIGARCH Computer Architiure News 37.2* (2009), pp. 46–55. ISSN: 0163-5964. https://doi.org/10.1145/1577129.1577137

The manufacturer's authorised representative in the EU is Springer Nature Customer Service Centre GmbH, Europaplatz 3, 69115 Heidelberg, Germany. If you have any concerns regarding our products, please contact ProductSafety@springernature.com

Printed and bound by CPI Group (UK) Ltd, Croydon, CR0 4YY
01/08/2026
02183526-0001